Alcimar das Chagas R. Ribeiro
Milton Erthal Júnior
Ana Paula Erthal

Instructions for growing Tilapia in plastic-lined tanks

Alcimar das Chagas R. Ribeiro
Milton Erthal Júnior
Ana Paula Erthal

Instructions for growing Tilapia in plastic-lined tanks

Fish farming

ScienciaScripts

Imprint

Any brand names and product names mentioned in this book are subject to trademark, brand or patent protection and are trademarks or registered trademarks of their respective holders. The use of brand names, product names, common names, trade names, product descriptions etc. even without a particular marking in this work is in no way to be construed to mean that such names may be regarded as unrestricted in respect of trademark and brand protection legislation and could thus be used by anyone.

Cover image: www.ingimage.com

This book is a translation from the original published under ISBN 978-3-330-99559-8.

Publisher:
Sciencia Scripts
is a trademark of
Dodo Books Indian Ocean Ltd. and OmniScriptum S.R.L publishing group

120 High Road, East Finchley, London, N2 9ED, United Kingdom
Str. Armeneasca 28/1, office 1, Chisinau MD-2012, Republic of Moldova, Europe
Managing Directors: Ieva Konstantinova, Victoria Ursu
info@omniscriptum.com

Printed at: see last page
ISBN: 978-620-3-31460-1

Presentation

Brazil has great potential for fish farming, as it has plenty of quality water, a suitable climate for fish to thrive for most of the year and relatively low-cost feed due to the wide availability of agricultural products.

In the coastal municipalities of the Norte Fluminense and Baixada Litorânea regions, this situation is no different, as some local climatic factors, such as high temperatures, sunshine and frequent winds, are favourable to fish farming. In addition, these municipalities: São Francisco do Itabapoana, São João da Barra, Campos dos Goytacazes, Quissamã, Carapebus, Macaé, Rio das Ostras, Casemiro de Abreu and Cabo Frio, have a tradition of freshwater and saltwater fishing, and this activity is one of the main sources of income for local families, so projects linked to fish farming present a favourable environment for development. One local restriction to building ponds for intensive fish farming on the coastline of these municipalities is the fact that the soil is highly sandy, favouring a high rate of water infiltration. However, a viable solution to this problem is to line the pond with plastic suitable for this purpose, which, despite increasing the cost of construction, provides satisfactory results for producers who use this system.

Despite the difficulties in obtaining statistics for the sector, in recent years there has been a decline in the fishing industry, with a reduction in marine shoals, which account for the largest share of fish, for example in the municipality of São João da Barra-RJ. Freshwater fishing, which is predominantly done in the Paraíba do Sul river, has also been declining over the years, and suffered a major impact in April 2003 with the ecological accident that seriously jeopardised the survival and quality of fish in this river. These facts contribute to the impoverishment of families linked to this activity, so some alternatives must be devised to improve income for fishermen's families and small rural landowners.

In this context, fish farming has emerged as an interesting alternative for generating business for fishing families and small rural landowners located along the coastline of the municipalities in the north of the state of Rio de Janeiro.

SUMMARY

CHAPTER 1

Alcimar das Chagas Ribeiro

Milton Erthal

Ana Paula Ribeiro Costa Erthal

1. introduction

Faced with the difficulties imposed on capture fisheries, such as heavy pollution in the seas, rivers and lagoons, as well as overfishing, pressurised by unemployment and deepening poverty in coastal regions, aquaculture has presented itself as an important business and food alternative for fishing populations.

According to the Food and Agriculture Organisation of the United Nations (FAO), aquaculture has seen significant growth around the world, compared to the development of capture fisheries. Figure 1 below shows the production trajectory of capture fisheries and continental and maritime aquaculture in the world.

Figure 1: The global state of fisheries and aquaculture
Source: Food and Agriculture Organisation of the United Nations - FAO (2016)

We can see that the volume of aquaculture fishing, which was 61.7 per cent of the volume of capture fishing in 2009, rose to 79.0 per cent in 2014. Following the same trajectory, per capita consumption in kg also grew over the period, from 18.1 kg in 2009 to 20.1 kg in 2014.

In Brazil, aquaculture is also making great strides due to the environmental imbalances that make it difficult to exploit capture fisheries. Statistics are scarce on production and producers of both types in the country. According to the FAO (2016), scale fish production totalled 474.3 million tonnes, while total production totalled 562.5 million tonnes in aquaculture in 2014. This figure is equivalent to 0.56 per cent of world production.

As we have seen, the development of aquaculture represents an important trend, given the problems related to capture fisheries. Its relevance is further emphasised when it is identified as a fundamental alternative to artisanal fishing. This important activity, due to its real decline, throws a large contingent of workers and their families to the margins of poverty and misery.

Below we present some aspects of artisanal fishing, a very important activity in terms of generating employment, income and food for people all over the world.

1.1 *Artisanal fishing*

Artisanal fishing is mainly characterised by family labour. It operates close to the coast, in rivers and lagoons. The equipment varies according to the species and can be classified as (seine, gillnet, single trawl, double trawl, tarrafa, traps, line and hook, among others). It is carried out solely and exclusively by the fisherman's manual labour. Man participates fully, or almost fully, in all stages of handling the implements and the product, without the need for mechanical traction when casting, retrieving and lifting nets or other implements.

Based on the knowledge passed down to fishermen by their ancestors, by the elders of the community or through interaction with fellow fishermen, artisanal fishing is always carried out in small boats (dinghies, dinghies and canoes), rowing or sailing or even motorised, with no instruments to support navigation, relying solely on experience, acquired knowledge and the ability to observe the stars, winds and tides. It does not rely on large-scale production or stockpiling.

Another important aspect observed is the fragility of this class of workers in relation to the suppliers of products and services and the intermediaries who buy their catch. Fishermen are the economically weakest part of the process and therefore have to bear high costs when buying inputs and settle for a selling price for their labour product that is far below the market price imposed on them by the buyer.

1.2 *Artisanal fishing in Brazil*

Brazilian artisanal fishing is a way of life for coastal populations. According to IBAMA, it accounts for around 55% of the fish consumed domestically. This activity, which is of fundamental importance in the process of generating work and income, depends fundamentally on public policies, which are rare and almost always inefficient. In this context, there are many problems. The precariousness of statistics can be considered a crucial problem. Without real knowledge of the activity, any public policy planning, as well as important decisions, are hampered.

In light of all the difficulties, according to estimates made by IBAMA in various publications, there are around 800,000 fishermen who make their living directly from fishing in Brazil, which represents an involvement of two million people, in a country of continental dimensions, but with strong regional differences.

These fishermen were "theoretically" organised into fishing **colonies** created **in** 1919 by the Navy. In reality, the aim of creating these organisations was to be able to link fishermen and fishing communities for possible war needs. The current observation is that these organisations do not make an effective effort to organise the fishing class and, more often than not, are involved with other interests that do not contribute to improving the living conditions of these individuals. It is estimated that there may be 600 fishing colonies in Brazil.

With regard to the volume of fish produced in the country, according to the IBGE and IBAMA, statistics show that one million tonnes of fish are produced each year, distributed as follows: 24.5% from industrial fishing; 50.1% from artisanal fishing and 25% from aquaculture.

The aquaculture activity, mainly fish farming, is the subject of this book and the location of the experience is the municipality of São João da Barra, located in the northern region of the state of Rio de Janeiro. Below we will present some aspects of artisanal fishing in the municipality and then we will deal specifically with fish farming.

1.3 *Characterisation of artisanal fishing in São João da Barra*

Fishing in the municipality of São João da Barra is artisanal and operates with approximately three hundred and fifty boats with an average length of twelve metres and an individual capacity of approximately four tonnes. The fleet has a fairly advanced average age, perhaps close to that of the Brazilian fleet, i.e. around twenty years old. The workforce is made up of approximately 1,500 professional fishermen who work in the fishing methods of fall fishing, shrimp fishing, platform fishing and Peruvian fishing. Completing this structure is a group of five shipyards that carry out their activities by hand. Four of these shipyards build boats and one only provides general repair services. Other complementary businesses, such as the supply of inputs to the boats, the timber trade, net repair services and small shops selling equipment in general are structured around this productive configuration in order to keep it functioning.

Fishing in the municipality is of fundamental importance to the population in many ways. Historically, the activity is intertwined with the colonisation process itself, which began in 1622. In that year, a group of fishermen from Cabo Frio found a favourable environment to develop their activity (OSCAR, 1999; SÁ, 1999).

According to Oscar (op. cit.), São João da Barra is the only city in Rio de Janeiro founded exclusively by fishermen. In this way, it can be said that this is a business that is crystallised in the culture of the people themselves. As in other coastal regions of Brazil, this enterprise is much more than an economic activity. It can be considered the materialisation of a way of life deeply rooted in the long history of coastal settlements. This ancient activity, represented by the interaction between natural resources and the skills developed by those who live along the coast, has made it possible for countless families to survive.

1.4 A brief account of the fish farming experience in São João da Barra

The municipality of São João da Barra, on the north coast of the state of Rio de Janeiro, has experimented with fish farming for tilapia through a pilot project financed by Petrobras under the Federal Government's Zero Hunger programme. The project arose for the same reasons as the rest of the country, i.e. heavy pressure on fish stocks in capture fisheries and environmental imbalance due to high levels of pollution (RIBEIRO and BODEN, 2012).

The project was conceived as an integrated system for fattening tilapia in small plastic-lined tanks suitable for sandy soils. The aim was to generate and systematise information on the various stages of handling, processing, storage and marketing, so as to be able to structure an information system on the business for the subsequent transfer of technical knowledge to interested rural producers in the region. The multiplication of the business was desirable due to its capacity to generate employment and income, and its chain-structured production organisation was essential for it to be competitive.

1.4.1 WHY TILAPIA?

Tilapia are fish of African origin, originally from the Ivory Coast, of the genus Oreochromis, which currently has around 70 described species. It was introduced to north-eastern Brazil in 1971 and then spread throughout the country, being cultivated from the Amazon river basin to Rio Grande de Sul. Interest in the cultivation of this species in the south and south-west of the country has grown rapidly in recent years due to the introduction of sex reversal technology and sport fishing. It is estimated that in Brazil, half of the annual production of fish by fish farms is tilapia (LOVSHIN and CIRYNO, 1998).

These fish have some characteristics that are highly favourable for breeding (KUBITZA, 2000), such as:

Adaptation to environmental conditions, they prefer temperatures around 25 ºC;

ˣˈ?ˮ Good feed conversion and low protein requirements, reducing production costs;

ˣˮ Rapid weight gain, allowing producers to harvest two or more crops a year;

ˣˈˀ⁵ Rusticity;

ˣˈ-ˀˈ Omnivorous feeding habits (feeds on a variety of sources), including plankton, which saves on feed;

ˈˣ-^ Adapts to confinement;

Resistant to low concentrations of dissolved oxygen in the water;

ˣ-ˀˈ Good acceptance on the consumer market, which means it's easy to sell due to its affordable price, pleasant flavour and low fat content;

ˣ-^ˈ Resistance to some diseases, which makes them one of the scale fish species with great potential for intensive farming.

Tilapia have great market potential, because in addition to fillet, which is in great demand both domestically and abroad (CREPALDI et. al., 2006), producers or slaughterhouses can increase their profit margin by processing the by-products of this fish, as shown in figure 2.

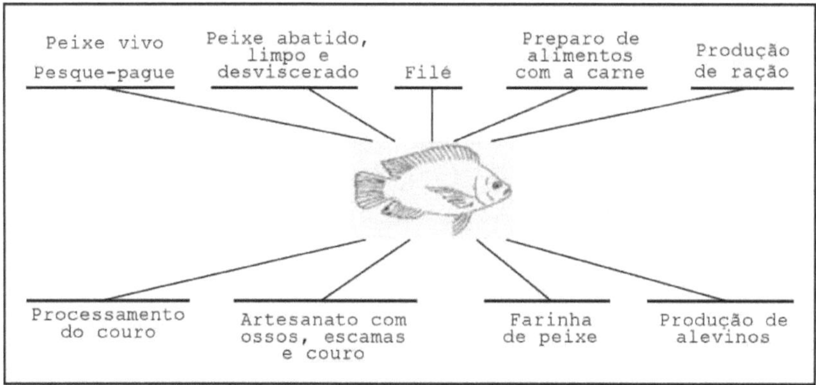

Figure 2 - Tilapia business and marketing possibilities.

The meat residues that adhere to the carcass after the filleting process can be removed and used to make hamburgers, sausages and other food products. Carcass meal, or fish meal, is an input that is widely absorbed by companies that produce fish feed, as this product contains compounds that are essential for fish nutrition. The processed skin is a by-product that can become more profitable than selling the fillet. Another option is to use the scales, bones and leather to make handicrafts. These parameters together point to a low risk of failure for the entrepreneur who allocates his resources to tilapia production or processing.

1.5 Methodological aspects

The process evolved on the basis of Action Research methodology, defined as empirically based social research, designed and carried out in close association with an action or the resolution of a collective problem, a condition in which researchers and participants who are representative of the situation or the

problem are involved in a co-operative or participatory way (THIOLLENT, 2002).

According to the author, the main intrinsic aspects of action research are: (a) there is a broad and explicit interaction between researchers and the people involved in the situation being investigated; (b) this interaction results in the prioritisation of the problems to be investigated and the solutions to be put forward in the form of correct action; (c) the object of the situation is not the people but the social situation and the problems of different kinds encountered in this situation; (d) the aim of action research is to solve or at least clarify the problems of the observed situation; (e) during the process, decisions, actions and all the intentional activity of the actors in the situation are monitored; (f) the research is not limited to one form of action (risk of activism): it is intended to increase the knowledge of the researchers and the knowledge or "level of awareness" of the people and groups considered.

These elements were implemented, with some adaptations due to a lack of resources. It took seven years (2004 to 2010) of rich experience, in which the project proved to be extremely viable, but political reasons made it impossible to continue, despite the fact that it left behind important knowledge related to its practice.

The end of the same experiment also left great expectations and individualised initiatives that need to be supported if they are to be successful. We believe that the revival of this experience, starting with the construction and operation of a demonstration plant, implemented with technical rigour and considering the organisation of the activity from the perspective of collective action, could represent an excellent business alternative with a high standard of employment and income, as well as a guarantee of social inclusion.

References

ASSUMPÇÃO, J; NASCIMENTO, MT. Structure and floristic composition of four restinga vegetation formations in the Grussaí/Iquipari lagoon complex, São João da Barra, RJ, Brazil. **Acta Botanica Brasilica 14:** 301-315. 2000.

CAPUCCI, E. **et al.** Tubular wells and other groundwater abstractions. DRM-RJDepartamento de Recursos Minerais do Rio de Janeiro, 2001.

CREPALDI, DV; FARIA, PMC; TEIXEIRA, EA; RIBEIRO, LP; COSTA, AAP; DE MELO, DC; CINTRA, APR; PRADO, AS; COSTA, FAA; DRUMOND, ML; LOPES, VE; MORAES, VE. The situation of aquaculture and fishing in Brazil and worldwide. Rev. Bras. Reprod. Anim. 30(3/4), 81-85, 2006.

FAO "Food and Agriculture Organisation of the United Nations" The State of Fisheries and Aquaculture. The Fisheries and Aquaculture Department's flagship publication issued every two years, provides a comprehensive, objective and global view of capture fisheries and aquaculture, 2010.

KUBITZA, F. Tilapia: production technology and planning. Jundiaí: Edcopyright, 2000. 289 p.

OSCAR, J. Notes on the history of São João da Barra. **São João da Barra Newspaper**, n. 46, p. 01-08, September 1999.

SÁ, CAA. Notes on the history of São João da Barra. **São João da Barra Newspaper**, n. 46, p. 01-08, September 1999.

CHAPTER 2

FRY PRODUCTION AND TRANSPORT

George Shigueki Yasui
Eduardo Shimoda

2.1. SIRE SELECTION

In order to start reproduction, it is interesting to plan the system to be adopted (egg or post-larvae collection), so that the matrices can be selected. This is because, as will be described later, in order to collect eggs from the mouths of the females, it is recommended to use smaller animals.

In general, females weighing 450 grams or more are used. Very young females tend to spawn less frequently and have a low number of eggs, as well as being smaller (GUNASEKERA et al., 1996a and 1996b; GOMEZ-MARQUES et al., 2003) - with less vitelline, which reduces the survival of the larvae.

On the other hand, very old females also tend to produce a low quantity of eggs, so it is suggested that they be discarded. It is recommended to discard females that are around 2 years old, or earlier, if there is a low productivity of larvae in a given batch (NANDLAL and PICKERING, 2004).

When matrices are stocked for reproduction, not all the females will reproduce at the same time, which generates post-larvae of different sizes and ages, jeopardising the effectiveness of sexual inversion. In addition, tilapia can practice cannibalism, especially in the early stages, which reduces the availability of post-larvae (PANTASTICO et al., 1988).

Given these facts, the best results occur when there is reproductive synchronisation, which generates a large number of larvae that are standardised in terms of age and size. An interesting method for achieving this standardisation is synchronising reproduction and stocking matrices that are at similar stages of gonadal development, so that the spawning period between the females is close.

According to the WORLD FISH CENTER (2004), standardisation can be seen in the external characteristics of the papilla and the volume of the abdomen (table 1). It should be borne in mind that these values can vary according to the conditions of the environment (temperature, food available, management), but they are good references for synchronising reproduction.

Table 1. External characteristics of female tilapia for predicting the day of reproduction

Category	External features	Days for reproduction
Spawning imminent	Bulky abdomen; prominent (open) papillae, pink or reddish in colour	3-7
Ready to spawn	Abdomen less voluminous; papilla pink or yellowish in colour and not very open	5-10
Not ready for spawning	Papilla light-coloured, retracted; normal abdomen	21-30
Spawned	Red coloured papillae; flaccid abdomen	15-30

There are numerous factors that can affect reproduction:

- Nutritional status of breeding stock
- Water quality
- Strains or species used
- Age and size of matrices

Before stocking for reproduction, it is recommended that the females are separated and fed good quality feed for a period of 15 days, and only after this period are the fish stocked for reproduction. This improves the condition of the animals and optimises the production of post-larvae. Male animals generally don't need this kind of treatment and can stay in the breeding tanks intermittently. For this pre-breeding phase, the animals can be stocked at a density of 1 animal per square metre, with periodic water renewal. If possible, the producer should have at least two breeding batches: one active batch (breeding) and one recovering batch. This avoids excessive wear and tear on the breeding stock.

Tilapia reproduction can be achieved all year round, as long as the water temperature is above 22°C (NANDLAL AND PICKERING, 2004), and best performance is observed at temperatures in the 28-30°C range. Breeders should be fed 1-3% of their live weight, consisting of a feed containing at least 30% crude protein (GUANASEKERA et al., 1996a).

Breeding ponds can be excavated with an earthen or masonry bottom, or even net ponds with a 1.5mm mesh, where larvae or even eggs can be collected.

In order to maintain a good breeding stock, we recommend:

- Obtain animals of proven genetic quality
- Maintain the genetic quality of the batch, avoiding contamination with other fish (control batches, crossbreeding, age, etc.)
- Eliminate fish with undesirable characteristics (poor performance, deformities, diseases that could jeopardise performance)
- Avoid inbreeding (crossing between close relatives). To do this, you should keep a larger herd and reduce the likelihood of crosses between relatives
- Periodic replacement of matrices

2.2. TILAPIA REPRODUCTION AND OBTAINING POST-LARVAE

Fish develop in several stages. After the eggs hatch, larvae are obtained, which have a nutritional reserve called the yolk sac. This reserve, which sustains the fish in the early stages of life, is gradually consumed, after which the fish turn into post-larvae (Figure 7). They then need to obtain external food. This is when sexual inversion takes place. The procedures for obtaining these post-larvae are described below.

According to PHELPS and POPMA (2000), tilapia reproduction can be carried out in excavated ponds or in hapas. In the first case, the fish reproduce in the tanks and the post-larvae are partially or totally collected (Figure 1), and in the second case, the eggs are collected, artificially incubated and the post-larvae are obtained. The system adopted will depend on the type of structure available (laboratory, tanks), the quantity of post-larvae required at one time and the resources available.

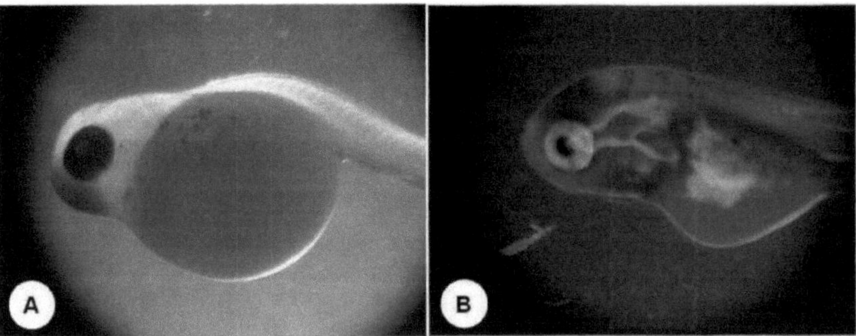

Figure 1 - Nile tilapia larva (**O. niloticus**), showing a large amount of vitelline reserve as its main characteristic (A); In contrast, post-larva of the same species, showing a scarce vitelline reserve (B).

2.2.1. Breeding in excavated ponds

A) For partial collection of post-larvae

The tanks used for the partial collection of post-larvae must be excavated, with an earthen or masonry bottom, since tilapia can reproduce **even without the presence of "nests". The size of the ponds should not be too** large, as this is detrimental to management, being around 200m^2, and they should not contain juvenile tilapia or any other species, as well as possible predators (dragonfly larvae - odonata, water cockroaches, etc.). For this reason, it is recommended that the pond be emptied, dried and levelled before stocking the matrices, and then limed (in the case of ponds with an earthen bottom). The matrices can be stocked at a ratio of one male to up to three females, with a stocking density of 300-700g/m^2. After 10 days of stocking the matrices, observations of post-larvae should be made in the marginal region of the tanks, preferably at the beginning of the day, on the brightest edges.

In this system, post-larvae are collected daily in the marginal region of the tanks using a paddle. The amount of post-larvae obtained per collection is around 1 post-larvae/kg female/month (PHELPS and POPMA, 2000). When daily collection is used, it has the advantage of requiring less management, being less stressful for the animals and the possibility of collecting for a longer period of time. As it is not possible to collect all the larvae in this method, the ones that remain in the tank can practice cannibalism.

Post-larvae can be collected using a fine-meshed (1.5mm), circular forceps with a long handle, which is passed with rapid movements close to the edge of the tank. To make it easier to handle, the mesh should be tensioned, **as opposed to traditional pincers where the mesh forms a sort of "bag".** The tensioned mesh allows for easier collection and the animals suffer less during handling. You can also build a square bag, measuring 1.5 x 1 metre, which can be handled by two people. The best time to observe and collect post-larvae is early in the morning, as temperatures are milder at this time and the animals suffer less from handling and heat stress. To avoid cannibalism, collections can be carried out for up to 40 days, when the tank should be emptied and prepared for a new cycle.

The collected animals are transferred to plastic buckets or styrofoam boxes, preferably containing water from the same pond where they were collected. However, if the water is very turbid or has low oxygen concentrations, it is advisable to use another water source to avoid mortality. Portable battery-operated aerators (used in aquariums) can also be used. The amount of water contained in this bucket varies depending on the ambient temperature (very hot days require larger amounts of water to slow down heating and the reduction of oxygen) and the number of larvae to be collected.

The size and age of the post-larvae are not standardised, so sorting is necessary to select only the animals of the right size (smaller than 14mm). Animals larger than 14mm can no longer be inverted, as the gonadal differentiation process is already advanced. Sorting can be done with the aid of a plastic picker (Figure 2), where only larvae up to 14mm can pass through. The plastic sifter can be obtained from specialised stores, or even improvised using pasta strainers, as long as the shape of the holes allows for this adaptation. In any case, the holes **must be "calibrated" with sandpaper** to adapt them to the size of the larvae.

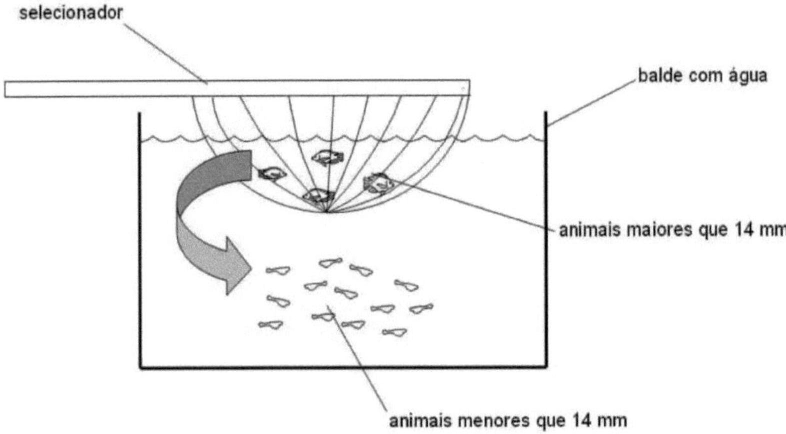

Figure 2: Screening of tilapia post-larvae to select animals for sex reversal

According to ZIMMERMANN (2000), "sorting" has the disadvantage of eliminating many animals with good growth potential and, as males grow faster than females, many of these are discarded. This tends to reduce the percentage of males at the end of the sex reversal process. However, with proper management, satisfactory sex reversal rates (95-99%) can be achieved.

B) For total collection of post-larvae

In the case of total harvesting, the entire tank is harvested by removing the matrices and collecting the post-larvae. The matrices can then be separated and the tank prepared for a new cycle. In this system, the matrices are stocked at a ratio of one male to two females, with a stocking density of approximately 500g matrices/m^2. The tanks must have a structure for collecting the animals (collection box), which should preferably be made of concrete, and have an area of 10m^2 or more, with a minimum of 1% of the tank's area, and a depth of between 30-40cm.

At the time of harvesting, the collection box is covered with a mesh that allows only the post-larvae to pass through, and the matrices are removed. The post-larvae can be collected in the collection box itself, using a fine-meshed square bucket, or at the water outlet of the tank, using a fine-meshed net. It is recommended that, before emptying and handling in the collection box, the water level in the tank is reduced to 50 per cent by removing part of the matrices using a coarser mesh trawl. This avoids an excess of animals in the collection box, which can increase post-larval mortality.

Mortality due to management is higher than with daily collection. The key point in using this system is the right time to collect the larvae, as in general there is no reproductive synchrony and the tank can often be emptied when many females are about to reproduce or are incubating eggs in their mouths. The post-larvae also need to be sorted by selecting the right size animals.

There are several equations in the literature for predicting the ideal day for management, mainly based on temperature. However, because there are several variables that can influence reproduction, such as feeding, the physiological state of the broodstock and the photoperiod (RIDHA and CRUZ, 2000; GUANASEKERA et al., 1996), the authors recommend daily sampling using a fine-meshed scoop to observe the presence of post-larvae. However, information on predicting the day of management can be found in Green and Teichert-Coddington (1994).

2.2.2. Breeding in hapas

Tilapia reproduction can also be carried out in net ponds or **"hapas"** (figure 11). The use of this type of tank for tilapia reproduction facilitates the collection of larvae, as well as making it possible to collect the eggs that are being incubated in the mouths of the females, so that they can then be incubated artificially. Another important aspect is that management in hapas allows spawning to take place in larger tanks where other fish are stocked. However, many matrices generate lower reproduction rates because they are not fully adapted to this type of confinement. Another disadvantage is the possibility of escapes if the mesh breaks, as well as the need to clean the mesh frequently to prevent it from clogging.The hapas used can be from 2 to >500m^2, with a mesh opening of 1.5mm, where the matrices are stocked, in a ratio of 1 male to up to two females. The number of matrices is around 400 to 600g of matrices per cubic metre, receiving feed similar to that described in excavated ponds. Eggs should be collected weekly, by scooping up all the females with a dipper and checking for the presence or absence of eggs in the oral cavity. To make handling easier, smaller females (around 500-600g) and larger males (around 700g) can be used. This makes it easier to collect the eggs, as smaller animals are easier to handle. The presence of larger males means that you don't have to waste time checking males for oral incubation, but marked males (plastic pieces, coloured wire, or spines removed from the dorsal fin) can also be used for this purpose. The procedure for collecting the eggs is described in figure 3. To facilitate handling, a floating divider is used, which can be made from bamboo, **with futuribles tied to the ends (e.g. pet bottles).**

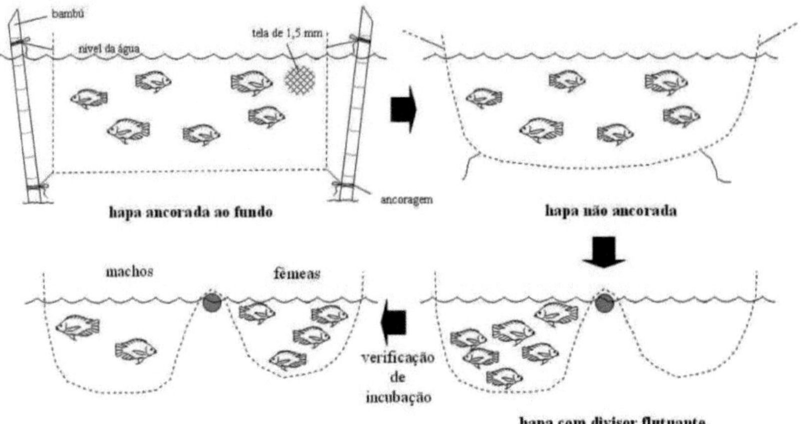

Figure 3 - Management adopted to check for oral incubation in tilapia kept in hapas

2.3. ARTIFICIAL INCUBATION

The technique of artificially incubating tilapia eggs is quite old (ROTHBARD and PRUGININ, 1975; RANA, 1988; HSIAO, 1977), however, in recent years it has been used commercially (ZIMMERMANN, 1999).There are various incubator models, however, the most commonly used is similar to the one described in figure 4.

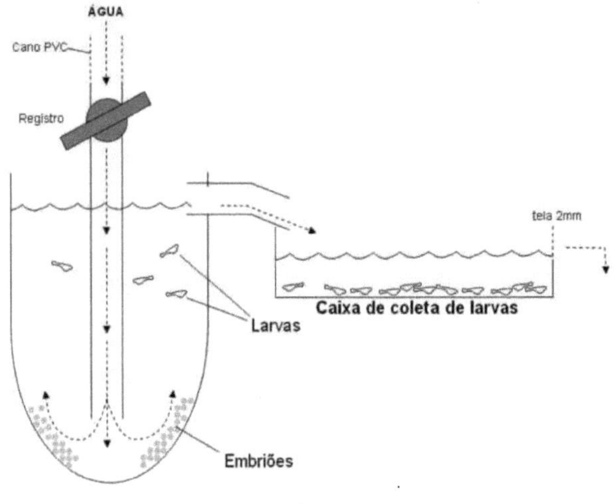

Figure 4 Artificial incubation system for tilapia eggs

The eggs are placed in the incubator and, because they are denser than the water, they stay at the bottom, being oxygenated by the water flow. When the eggs hatch (about 2-3 days), the hatching larvae go to the top, until they are sent to the larvae collection box. This collection box can be a plastic box with a 1.5mm screen at the outlet to allow the larvae to be collected, or it can be a small hapa (1.5mm opening) immersed in a larger tank or container.

The concave part of the incubator (Figure 4) can be constructed using chicken troughs, buckets or inverted plastic bottles (YASUI et al, 2002), filling **the "neck" part with plastic** filler or resin, or another material.

similar material. However, containers with low porosity (smooth) should be used so that the movement of the incubation does not jeopardise the quality of the eggs.

Before inserting the eggs into the incubator, unwanted particles such as pebbles, algae and scales should be removed using tweezers and/or a sieve. During incubation, the ideal water flow will basically depend on the number of eggs, the size of the incubator, the curvature of the concavity and the distance between the water inlet pipe and the bottom of the incubator. It is therefore recommended that the producer **finds the ideal flow rate for the incubator used by "trial and error".** As a general rule, the flow should be constant, oxygenating the eggs evenly and not moving them too vigorously. As the eggs develop, adjustments to the flow rate of the incubators may be necessary. A temperature of 26-28°C is recommended, with as little variation as possible. Incubation time varies between 2 and 3 days, depending on the stage of development at which the eggs were collected. Eggs at a more advanced stage of development are darker in colour.

The larvae can be collected daily and stored in another larger container (e.g. water tank or gutter), with constant water renewal, until they absorb the yolk sac. After absorption, the animals are counted and sent to the sex reversal process (as described above). When the eggs are incubated artificially, animals of a **standardised size and age are** obtained, **so "sorting" is not necessary.**

The hatching rate is generally high (>80%), but the strain used must be taken into account, as the eggs of some red tilapia strains have low hatchability (LOVSHIN, 2000). If mortality at this stage is very high, the eggs may be being attacked by fungi or bacteria (mainly). To overcome this problem, a 100ppm iodine solution can be used. To do this, switch off the water supply, add the iodine in the indicated proportion and leave the disinfectant to act for around 7 minutes. During this disinfection, oxygenation is necessary, which can be done by periodically turning over the water in the incubator itself, using a small container (e.g. a glass). After this period, water renewal is resumed, but care should be taken if the incubators are connected to a recirculation system, as iodine can jeopardise biological filtration.

2.4. SEXUAL INVERSION

The selected larvae can then be counted and sent to the sex reversal tanks. Counting can be done in two ways: for small quantities, it can be done using fine mesh sieves, counting the animals as quickly as possible and then sending them to the reversal tanks. When counting, it is recommended not to collect too many larvae in the sieve (maximum of 50), since until all the animals have been counted, they will spend a long time out of the water, which can increase mortality. However, if you want to count a very large number of animals, it is advisable to use the volumetric method.

This method consists of filling a container with post-larvae, counting the number of post larvae, and then estimating the number of larvae using this **container as a "measure" by extrapolation. The container should not** be too large in volume (maximum 5-10mL), and the surface area should not be too large either (container taller than it is wide), as this increases the accuracy of the count. It is recommended that at least three counts are made, then the average value is obtained, which will be used for extrapolation.

Stocking density in inversion tanks varies according to the type of **tank (masonry, water tank, hapas** - figure 5) and the quality of the water. Care must be taken with the water renewal rate, the amount of feed and the stocking density.

Figure 11. Hapas where post-larvae can be stocked for sex reversal

In the case of masonry tanks, a reference density of 0.7 to 1.5 larvae per litre can be adopted for tanks with a low water renewal rate. In net tanks, the reference density is around 3000 to 4000 larvae per cubic metre. However, these densities vary according to the type of management adopted and the quality of the water. Therefore, the producer will have to use personal experience to define the density that best suits his conditions.

2.4.1. Preparing feed and feeding the animals

To prepare the feed for sexual inversion, a feed containing 35 to 50 per cent crude protein is used, which is finely ground so that the animals can ingest it, using a 0.6mm sieve. Minerals and vitamins can be added to this feed, especially if the animals don't have access to natural food (plankton).

The synthetic hormone 17-a-methyl-testosterone is then added to provide sexual reversal. This hormone can be purchased from specialised companies.

17-a-methyl-testosterone is insoluble in water, but is soluble in ethyl alcohol (clean alcohol - at least 90°GL), and can be degraded by light and high temperatures. To facilitate the preparation of the masculinising feed, a stock solution is prepared by dissolving 6g of the hormone in a litre of ethyl alcohol (unflavoured). This solution can be stored, preferably in a glass container, covered with aluminium foil and kept refrigerated.

The summary for preparing the feed is shown in figure 6.

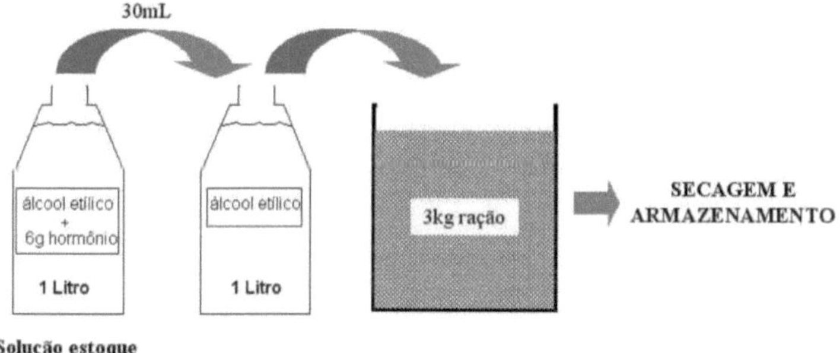

Figure 6 - Scheme for preparing sex reversal feed

To prepare three kilos of feed, 30 mL of this stock solution is used, which is then added to one litre of ethyl alcohol and mixed well. The resulting solution is added to three kilos of feed, mixing evenly. This should preferably be done in a glass container to prevent the container from retaining the hormone, although plastic containers with low porosity can also be used. After mixing, the feed should be dried in the shade, spread out on smooth plastic, in a ventilated environment. After drying, the feed should be stored in plastic bags and preferably kept refrigerated. It can be used for up to 60 days without compromising the effectiveness of sex reversal (SMITH and PHELPS, 2001).

2.4.2. Feed supply

Feed can be offered two to four times a day (preferably four), and the amount varies depending on the biomass of the tanks. In the first week of the inversion treatment, about 25 per cent of the biomass (total weight of the population) of the tanks is fed per day, divided into daily meals. In the following weeks, the amount of feed supplied varies according to Table 2.

Table 2. Percentage of biomass to quantify the masculinising feed

Week	Percentage of biomass*
1	25%
2	15%
3	10%
4	10%

To calculate the biomass, a sample is taken of a few animals (around 70), the total length of which is noted down and the average value obtained. From this value, the average biomass can be obtained, according to the equation described by POPMA and GREEN (1990):

Biomass of 1000 animals = 0.02 x (average length, in mm)^3

The daily amount of feed for 1000 animals will be:

Amount of food = Biomass of 1000 animals x percentage of biomass* Here is an example of a tank in the second week of sexual inversion: **Average sample length:** 15mm

Biomass of 1000 animals: $0.02 \times (15)^3 = 67.5g$

Total population: 4000 fry per tank

Total biomass: **67.5g x 4 = 270g**

Daily ration: **270g x 0.15 = 40.5g**

Ration per meal (3 meals a day): **40.5g/3=13.5**g

The animals should be sampled weekly to update the amount of feed. The feed should be spread evenly over the surface of the tanks to make food available to all individuals.

The sex reversal process lasts around 28 to 30 days and approximately 300 to 450g of feed is used for every 1,000 reversed fry, with a male percentage of 95-99% expected. At the end of the reversal period, the animals are collected, counted and the smaller ones removed, as they have consumed a small amount of feed and there is a high probability that they are female.

The animals should be sent to rearing ponds, and can be fed a diet designed for fry. The rearing pond can also be fertilised so that the animals can take advantage of the natural feed, reducing feed costs. After this period, when the fry are around 80 grams, the fish can be transferred to fattening ponds. Both newly hatched fry and juveniles can be marketed.

2.5. TRANSPORT OF FRY

Transport is a routine practice, according to Carmichel et al (2001) and of fundamental importance in fish farming systems, as most Brazilian fish farmers do not produce fry on their property and these are transported considerable distances. For this reason, it is routine for fish farms that produce fry to use predetermined quantities of fish in plastic bags, taking into account the following factors: the size of the fry, the species and the volume of water added. In addition, according to Malta (2006), the size of the bags, the amount of water and fry is highly variable, depending on the species, the estimated period of transport and the water temperature.

During transport, fish are affected by a series of stressors: capture, confinement, handling and transport itself (GOMES et al., 2006). Therefore, when transporting fish, special care must be taken to avoid problems with mortality (DIAS; MARUYAMA, 2006).

According to Dias and Maruyama (2006), it is recommended that the fish and fry fast for one or two days before being subjected to the transport conditions, during which time the animals' digestive tracts will be emptied, reducing the amount of waste in the transport water, as the stress caused by capture causes the fed fish to eliminate excrement several times. Fish droppings and excreta are known to greatly worsen water quality, reducing oxygen levels (BALDISSEROTTO, 2005).

Fish that are not fasted consume more oxygen and excrete more carbon dioxide and ammonia than fish that are fasted before transport. Several studies have also shown the presence of various pathogenic bacteria in the intestinal flora of fish. When fish are transported without fasting, the faeces excreted in the packaging or transport box inoculate the water with these pathogenic bacteria. In transport water loaded with organic waste (faeces, fish mucus), these bacteria multiply rapidly, increasing the potential for fish infection and transport losses (KUBITZA, 2003).

Density is also a special factor to be respected because, according to Dias and Maruyama (2006), high density, with stressed fish, can lead to cannibalism, where fish often tear off part of the caudal fin and bite the entire body of others, losing a lot of mucus through friction and causing a series of problems. The loss of mucus causes opportunistic pathogens such as fungi and bacteria to attack the lesions, often causing death or wounds that are difficult to recover from. A standard density of 50 fry per 3 litres of water is recommended by Dias and Maruyama (2006), or, according to Carvalho (2007), 50 grams of fish per litre of water can be used.

Another factor that can be observed to prevent stress during transport is the time at which it is carried out . It is recommended to transport fish in the early hours of the day or late afternoon, when temperatures are milder, avoiding the hours between 10am and 6pm (OKUMURA, 2005). If the water temperature rises, the fish's metabolism and activity accelerate, resulting in greater oxygen consumption and excretion of toxic metabolites (ammonia and carbon dioxide), as well as faecal waste (KUBITZA, 1998). All of this also contributes to the organism's stress process, which can lead to the death of the fish (OKUMURA, 2005).

According to Malta (2006), resistant plastic bags filled with water and pure oxygen injected under pressure should be used. The bags are closed using elastic bands (such as tyre alloys) to ensure secure and efficient tying (OKUMURA, 2005). The proportion of oxygen and water according to Okumura (2005), reinforced by Dias and Maruyama (2006), is 2/3 and 1/3 respectively.

Prophylactic measures such as the use of anaesthetics and/or salt in the water can also be used to reduce stress. According to Dias and Maruyama (2006), it is recommended to introduce salt into the transport water at a rate of 3 to 5 grams of salt per litre of water.

Table salt as a stress reducer can be used in aquaculture to equalise the osmotic gradient between the water and the fish's plasma, reducing the diffusion of ions into the water. Salt also stimulates the secretion of mucus on the gill epithelium, diffusing the passage of ions through the cell membranes (GOMES et al., 2003), helping to cover up injuries sustained during the handling of fish, and can help control some parasites during transport (KUBITZA, 2003).

According to KUBITZA (2003) the use of anaesthetics is particularly useful when loading juvenile fish, especially those that are more agitated, such as piauçu, dourado, black - bass and piraputanga. Sedation (a mild effect caused at a certain stage of anaesthesia, in which the senses are impaired and there is little or no recognition of pain) reduces the metabolic rate and consequently oxygen consumption and the excretion of metabolic products (ALVES et al., 2001). According to ROUBACH & GOMES (2001); IVERSEN et al. (2003); PIRHONEN & SCHRECK, (2003) there are dozens of anaesthetics used in fish farming, including Benzocaine, Menthol, clove oil, MS-222 (tricaine methane sulfanate) and Phenoxyethanol. Lavor et al. (2007) and Sartore (2007) studied the optimum concentration of benzocaine and menthol, respectively, for piauçu fry.

As for the procedure adopted when the fry arrive at the destination property, the fish should spend a period of approximately 20 minutes in the transport container with the container still sealed in the water of the pond they will be introduced into, so that there is a balance in the temperature of both containers (and consequently of the animal's body), avoiding shocks thermal. In general, the transport water is warmer than the pond water. Once this equilibrium period has passed, the plastic bags are opened, allowing the water from the pond to mix with the water in the bag, and the fish slowly leave the container (OKUMURA 2005).

According to Grottum et al. (1997), the main factor for successful transport is to contain the highest density of fish in the smallest possible volume of water, without any mortality, deterioration in water quality or stress. The ideal density for transporting piauçu fry was studied by Pessôa (2007). This author found that the longer the transport time, the lower the permitted density, in order to minimise mortality.

Fry and post-larvae can be sold as long as they are in good sanitary condition and are packed in plastic bags with oxygen under pressure. Before being packaged, the fish must be left for a period without food, in running water, to empty their intestines and prevent the elimination of faeces in the packaging from spoiling the water. This period varies depending on the size of the fry and the temperature. The conditioning time (fasting) and the number of fry that can be transported in plastic bags are shown in Tables 3 and 4, respectively.

TABLE 3: Fasting time (in hours) of fry before transport.

Fry length (cm)	Fasting hours
2-3	6-9
3-5	10-15
above 5	24

TABLE 4: Number of fry transportable in plastic bags (in 12 litres of water with 30 litres of oxygen, at 24-28°C)

Transport time (hours)	Fry length (cm)			
	2-3	3-5	5-8	8-10
6	3500	3000	2000	1500
12	3000	2500	1500	1000
24	2500	1500	800	400
36	1800	1000	600	300
48	1500	800	400	200

<u>REFERENCES</u>

ANDRADE, D.R.; YASUI, G.S. 2003. Management of natural and artificial reproduction and its importance in fish production in Brazil. Brazilian Journal of Animal Reproduction, v. 27, n. 2, p. 166-172

BAROILLER, J.F.; GUIGEN, Y.; FOSTIER, A. 1999. Endocrine and environmental aspects of sex differentiation in fish. Cellular Molecular Life Sciences, v.55, p.910- 931,

BOCEK, A. 2005. Culture of Hand-Selected Male Tilapia. International Centre for Aquaculture and Aquatic Environments, Auburn University, Alabama 36849-5419 USA. Available at: http://www.ag.auburn.edu/fish/icaae/publications.htm. Accessed in October 2005.

CURTIS, L.R., DIREN, F.T.; HURLEY, M.D.; SEIMAND, W.K.; TUBB R.A.. 1991. Disposition and elimination of 17"-methyltestosterone in Nile tilapia, (**Oreochromis** *niloticus).* Aquaculture, v. 99, p.193-201.

FISHBASE. 2005. www.fishbase.org. Accessed July 2005

GREEN, B.W.; TEICHERT-CODDINGTON. D.R. 1994. Growth of control and androgen-treated Nile tilapia, *Oreochromis niloticus* (L.), during treatment, nursery and grow-out phases in tropical fish ponds. Aquaculture and Fisheries Management 25:613-621.

GUNASEKERA, R.M.; SHIM K. F.; LAM, T. J. 1996a. Influence of protein content of broodstock diets on larval quality and performance in Nile tilapia, *Oreochromis niloticus* (L.) Aquaculture, v.146(3-4), p. 245-259

GUNASEKERA, R.M.; SHIM K. F.; LAM, T. J. 1996b Effect of dietary protein level on spawning performance and amino acid composition of eggs of Nile tilapia, Oreochromis niloticus . Aquaculture, v.146(1-2), p.121-134

GÓMEZ-MÁRQUEZ, J.L., PENA-MENDOZA1, B.; SALGADO-UGARTE, I.H.; GUZMÁN-ARROYO, M. 2003. Reproductive aspects of *Oreochromis niloticus* (Perciformes: Cichlidae) at Coatetelco lake, Morelos, Mexico Rev. Biol. Trop. n. 51, v.1, p. 221-228

GUERRERO, R.D.; SHELTON, W.L., 1974.An aceto-carmine squash method for sexing juvenile fishes. Progr. Fish-Cult., Bethesda, v. 36, n.1, p.56

JOHNSTONE, R., D.J. MACINTOSH AND R.S WRIGHT. 1983. Elimination of orally administered 17"-methyltestosterone by Oreochromis mossambicus (Tilapia) and Salmo gairdneri (Rainbow trout) juveniles. Aquaculture, v.35, p.249-257.

HSIAO, S.M. 1977. Artificial incubation of Tilapia eggs. China Fish. Mon., v.294, p.14-15.

KIM, D.S.; NAM, Y.K.; JO, J.Y. 1997. Effect of oestradiol-17B immersion treatments on sex reversal of mud loach, *Misgurnus mizolepis* (Gunther). Aquaculture Research, V. 28(12), p. 941-946

MAIR, G.C., J.S.ABUCAY, J.A. BEARDMORE AND D.O.F. SKIBINSKI. 1995. Growth performance trials of genetically male tilapia (GMT) derived from YY males in *Oreochromis niloticus* L.: On station comparisons with mixed sex and sex reversed male populations. Aquaculture, v. 137, p.313-322.

LOVSHIN, L.L. 1976. Summary of methods for growing tilapia. DNOCS, Fortaleza, 5p.

LOVSHIN L.L. 2000. Criteria for selecting Nile tilapia and red tilapia for culture. In: Tilapia Aquaculture in the 21st Century. Proceedings from the Fifth International Symposium on Tilapia in Aquaculture,Vol. 1, Rio de Janeiro, Brazil (ed. by K. Fitzsimmons & J. Carvalho Filho), pp. 49-57. Ministry of Agriculture, Rio de Janeiro, Brazil.

NANDLAL, S., PICKERING, T. 2004. Tilapia fish farming in Pacific Island countries. V. 1. Tilapia hatchery operation. Noumea, New Caledonia: Secretariat of the Pacific Community. 41p.

NOAKES, D.G.L.; E.K. BALON, 1982. Life histories of tilapias: an evolutionary perspective. In: INTERNATIONAL CONFERENCE ON THE BIOLOGY AND CULTURE OF TILAPIAS, 7., Manila, 1982. Proceedings. Manila: International Centre for Living Aquatic Resources Management, 1982. p. 61-82.

PANDIAN, T. J.; SHEELA, S. G., 1995. Hormonal induction of sex reversal in fish. Aquaculture, v.138, p.1-22.

PANTASTICO, J.B.; DANGILAN, M.M.A.; EGUIA, R.V. Cannibalism among different sizes of tilapia (*Oreochromis niloticus*) fry/fingerling and the effect of natural food, p.465-468. In:. The Second International Symposium on Tilapia in Aquaculture (Pullin, R.S.V.; Bhukaswan, T.; Tonguthai, K.; Maclean, J.L. eds.). Department of Fisheries, Bangkok, Thailand, and International Centre for Living Aquatic Resources Management, Manila, Philippines. ICLARM Comference Preoceedings 15, 1988, 623p.

PHELPS, R.P.; POPMA, T.J. Sex reversal of tilapia. In: COSTA-PIERCE, B.A.; RAKOCY, J.E (Eds.) Tilapia Aquaculture in the Americas. Louisiana: World Aquaculture Society, 2000. v.2, p.34-59.

PHILIPPART, J.C.; J.C. RUWET, J.C. Ecology and distribution of tilapias In: INTERNATIONAL CONFERENCE ON THE BIOLOGY AND CULTURE OF TILAPIAS, 7., Manila, 1982. Proceedings. Manila: International Centre for Living Aquatic Resources Management, 1982. p.15-61.

POPMA, T.J.; GREEN, B.W. 1990. Sex reversal of tilapia in earthen ponds: aquaculture production manual. Alabama: Auburn University. Research and Development Series, 35, 15p.

RANA, K. 1988. Reproductive biology and the hatchery rearing of tilapia eggs and fry. **In:** Muir, J.F., and Roberts, R.J. (Editors). Recent Advances in Aquaculture, Vol. 3, Timber Press, Portland, OR., U.S.A. p. 343-406.

RIDHA, M. T.; CRUZ, E M. 2000. Effect of light intensity and photoperiod on Nile tilapia **Oreochromis niloticus** L. seed production Aquaculture Research v. 31(7) p. 609

ROTHBARD, S.; PRUGININ, Y., 1975. Induced spawning and artificial incubation of Tilapia. Aquaculture, v.5, p.315-321.

SMITH, E.S.; PHELPS, R.P. 2001. Impact of feed storage conditions on growth and efficacy of sex reversal of Nile tilapia. North American Journal of Aquaculture, v.63, p.242-245

TREWAVAS, E., 1982a. Tilapias: Taxonomy and Speciation. In: INTERNATIONAL CONFERENCE ON THE BIOLOGY AND CULTURE OF TILAPIAS, 7, Manila, 1982. Proceedings. Manila: International Centre for Living Aquatic Resources Management, 1982. p. 3-14.

TREWAVAS, E., 1982b. Generic grouping of Tilapiini used in aquaculture. Aquaculture, v.27, p.79-81

VAZZOLER, A.E.A.M.. 1996. Reproductive biology of teleost fish: theory and practice. Maringá-PR: EDUEM. 169p.

VERA CRUZ, E.M.; MAIR, G.C. 1994. Conditions for effective androgen sex reversal in

Oreochromis niloticus, Aquaculture, v.122, p. 237-248,

WASSERMANN,G.J.; AFONSO, L.O.B.. 2002. Validation of the aceto-carmine technique for assessing the sex of nilotic tilapia fry (**Oreochromis niloticus**). Cienc. Rural, 2002, vol.32, n.1, p.133-139.

WORLDFISH CENTRE. 2004. GIFT Technology Manual: An aid to Tilapia selective breeding. WorldFish Centre, Penang, Malaysia, 56 p.

WOYNAROVICH, E.; HORVATH, L..1989. The artificial propagation of fish from tropical waters. Translation by Vera Lúcia Mixtro Chama. Brasília: FAO/CODEVASF/CNPq. 225p.

YASUI, G. S. ; NAVARRO, R. D. ; SANTOS, L. C. ; VIDAL JR., M. V. ; RIBEIRO FILHO, O. P. ; CALADO, L. L. ; BALBINO, E. M. 2002. Artificial incubation of Nile tilapia eggs using alternative incubators. In: XII Brazilian Aquaculture Symposium, 2002, Goiânia-GO. Proceedings of the XII Brazilian Aquaculture Symposium, 2002. p. 346-346.

ZIMMERMANN, S. 1999. Artificial incubation - a technique for producing genetically superior Nile tilapias. Panorama da Aquicultura magazine, Rio de Janeiro, v.9, n.4, p.15-21,

ZIMMERMANN, S. Good performance of chitraladas in Brazil. Revista Panorama da Aquicultura, Rio de Janeiro, v.10 n.60 p.15-19, 2000.

CHAPTER 3

WATER QUALITY IN TILAPIA FARMING

Milton Erthal Junior

Ana Paula Ribeiro Costa Erthal

Assessing water quality should be one of the first measures taken by anyone wishing to start fish farming. Studies by Capucci et. al. (2001) revealed the presence of underground aquifers with excellent water quality in the coastal region of Norte Fluminense. However, these wells are located at great depths, making it unfeasible to tap them unless they are subsidised for rural producers.

The groundwater in this region is shallow and can be collected cheaply using hydraulic pumps or wind turbines. Over the years, many rural producers have used this water to grow various species of fish without any major problems. However, when it comes to intensive cultivation in plastic-lined tanks, the quality of this water requires special attention, as preliminary observations suggest variations in pH, iron content, salinity, total ammonia and colour, often in wells that are close together. It is therefore prudent to send samples of the water to be used in fish farming for microbiological and physical-chemical analysis (MARTINS, 2007).

The waters of the Paraíba do Sul river, lakes and lagoons in the northern region of Rio de Janeiro have physico-chemical characteristics suitable for tilapia production, and the monitoring of certain parameters, such as heavy metal content, transparency, salinity and organic matter content, deserves more attention.

One of the reasons why tilapia is one of the most cultivated fish in the world is its tolerance to extreme conditions of some physical-chemical characteristics of the water that would be lethal for other fish species. Although there are differences between tilapia species and hybrids, they are generally tolerant of low levels of dissolved oxygen in the water, high levels of toxic ammonia, and adapt to a wide range of pH, reachability, salinity and temperature (KUBTIZA, 1999).

Next, the most important physical, chemical and biological factors of water for tilapia farming will be described.

3.1 - Physical properties

TEMPERATURE - is one of the most important factors for aquatic organisms, affecting the development of fish, phytoplankton and zooplankton. The thermal comfort range for tilapia species is between 25 and 32 °C (Table 1). Temperatures below 20 °C drastically reduce tilapia metabolism.

Temperatures above 32 °C cause thermal stress to the fish and affect water quality by decreasing the dissolved O_2 content and increasing the toxic ammonia (NH_3) content (KUBTIZA, 2003). Therefore, the risk of asphyxiation and ammonia poisoning on hot days is high, requiring extra attention from producers. In the Norte Fluminense region, the main problem is high water temperatures in summer, which can be controlled by renewing the water in the tank.

TRANSPARENCY - the colour of the water in plastic-lined cultivation tanks is generally turbid, due to the plankton. Transparent water not only indicates a shortage of algae, but also increases the risk of predation by birds. Green or blue-green water indicates good primary productivity (phytoplankton), which is desirable in tilapia farming because it is a natural food. The water temperature in disturbed ponds is generally higher than in ponds with clear water due to the absorption of heat by particulate material. Transparency can be determined using a Secchi disc or your own hand, which should be inserted vertically into the water. Ideally, the hand or disc should be visible at a depth of 20 cm (BOYD & TUCKER, 1998), because in very turbid or greenish water, dissolved oxygen can be compromised at night or on cloudy days. To minimise the turbidity of the water, simply stop fertilising and renew the water in the pond.

3.2 - Chemical properties

OXYGEN (O_2) - the main source of O_2 for pond water is algae, which produce it through photosynthesis. The diffusion of O_2 from the atmospheric air into the water is generally slow, except under conditions of strong turbulence such as wind. Tilapia are generally resistant to low concentrations of O_2 dissolved in the water, but below 3.0 mg O_2/l their development is affected (KUBTIZA, 2000). The concentration of oxygen in the water is affected by temperature, because the higher the temperature, the lower the O_2 concentration, and consequently the lower the biomass of fish supported, because the more fish, the greater the consumption. Another factor that affects temperature is the amount of plankton, because in the presence of the sun's rays, they photosynthesise and release O_2 into the water, but at night, algae also consume O_2 through respiration. This problem is exacerbated on cloudy days, as there is a shortage of sunlight, reducing O_2 production throughout the day. Suggestions for minimising the problem are described in Table 1.

AMMONIA - results from microbial action on organic waste (leftover feed, for example), fish excretion and other aquatic organisms. Ammonia is a gas that is very soluble in water and occurs in two forms: ionised (NH_4^+) and the non-ionised form (NH_3), which is the most toxic for fish. Kits for analysing ammonia in water generally quantify total ammonia ($NH_4^+ + NH_3$), and the percentage of toxic ammonia (NH_3) in relation to total ammonia increases with increasing temperature and pH (Table 2). Ideally, the level of toxic ammonia should be below 0.2 ppm or 0.2 mg/l (XXX). Tilapia tolerate NH3 values above the critical level well, but if they are constantly kept under this condition, intoxication can occur, the signs of which are: reduced feeding and growth, mouthing on the water surface during the day, incidence of disease and death of fish (Table 1) (PEREIRA & MERCADANTE, 2005).

Table 1. Response of tilapia to variations in physico-chemical parameters of cultivation water and procedures to resolve the causes.

Parameters	Tilapia responses	Procedures for resolving the causes of problems
Temperature (°C)		
8 a 14	Death.	Building a greenhouse over the tank or water heating system.
14 a 20	Susceptibility to disease, drastic reduction in appetite and low tolerance to transport and handling.	
20 a 25	Reduced appetite and growth.	
25 a 32	Ideal range.	
< 32	Reduced appetite and growth.	Renewing the pond water; building deeper ponds.
< 38	Death by heat stress.	
O₂ (mg/litre)		
0	Death after a few hours.	Activate an emergency aeration system; increase the frequency of water renewal; check the colour of the water and renew it if there is an excess of algae. Reduce the density of fish in the pond. Reduce feeding.
0 a 1,0	Death after a few days.	
1,0 a 2,5	Escape behaviour, darkening of the coypus, erection of the dorsal fin.	
> 3,0	Reduced activity.	
< 3,0	Ideal.	
Parameters	**Tilapia responses**	**Procedures for resolving the causes of problems**
PH		
2	Death in 12 hours	
3	Death in 1 to 3 days.	
4	40% mortality in 1 month.	
>4,5	Risk of death. Signs of asphyxiation accelerated opercular movements, mouthing of the surface,	

	increased mucus secretion, irritation and swelling of the gills. Dead with mouth open and eyes bulging.	
6,0 a 8,5	Ideal range	
< 10,5	Reduces feeding and development. Significant mortality after a few days.	
NH$_3$ (mg/1)		
6,6	50% mortality in 24 hours	Renew water, stop feeding.
4,0	50% mortality in 48 hours	
2,6	50% mortality in 96 hours	
0,2 a 2,6	Critical level. Reduced growth, feed conversion, tolerance to handling, transport and disease.	
> 0.2	Ideal	
Source: MARTINS, 2007; KUBTIZA, 1999, 2003; BOYD & TUCKER, 1998.		

Table 2. Percentage of NH3 (toxic ammonia) in relation to total ammonia as a function of pH and water temperature.

PH	Water temperature in °C					
	22	24	26	28	30	32
6,0	0,05	0,05	0,06	0,07	0,08	0,09
6,5	0,14	0,17	0,19	0,22	0,25	0,29
7,0	0,45	0,52	0,60	0,69	0,79	0,90
7,5	1,41	1,63	1,87	2,14	2,45	2,80
8,0	4,33	4,97	5,68	6,48	7,36	8,35
8,5	12,5	14,2	16,0	18,0	20,1	22,4
9,0	31,2	34,3	37,6	40,9	44,3	47,7

Source: KUBTIZA, 2003.

pH - Its scale ranges from 0 to 14 and the values are influenced by the concentration of H^+ (hydrogen ions) or OH^- (hydroxyl ions) in the water. Acidic waters (pH < 7.0) have an abundance of H^+ ; alkaline or basic waters (pH > 7.0) are rich in OH^-, while values around 7 characterise neutral waters. The water table in the coastal lowlands of São João da Barra has values of between 7.5 and 8.5, which is therefore within the ideal range for tilapia cultivation (ESTEVES, 1998). In ponds with low water renewal, the CO_2 dissolved in the water is abundant at night and low during the day and thus influences the pH of the water, which peaks in alkalinity and acidity at the end and beginning of the day, respectively. Alkaline values are problematic in relation to the ammonia dissolved in the water (Table 2). More information can be found in Table 1.

SALINITY - salinity is the total concentration of dissolved ions in water. It is a characteristic of water that is difficult to correct and is influenced by rain, which decreases the salt concentration, and evaporation, which increases the concentration of ions in the water. Ideally, salinity should be below 15 mg of salt per litre, but some species of tilapia, such as the red tilapia (Saint Peter) and **O. mossambicus** develop and reproduce in concentrations of 17 to 32 mg of salt per litre. Salt tolerance increases with the age of the fish, with fry being more sensitive to brackish water (ESTEVES, 1998).

Some of the factors discussed above, such as salinity and transparency, generally don't show sharp daily variations in their values, while the others fluctuate significantly over 24-hour periods. Figure 2 represents, in a generalised way and on an arbitrary scale, the fluctuation of pH, dissolved O2, ammonia, temperature, CO2, salinity and transparency values over a 24-hour period.

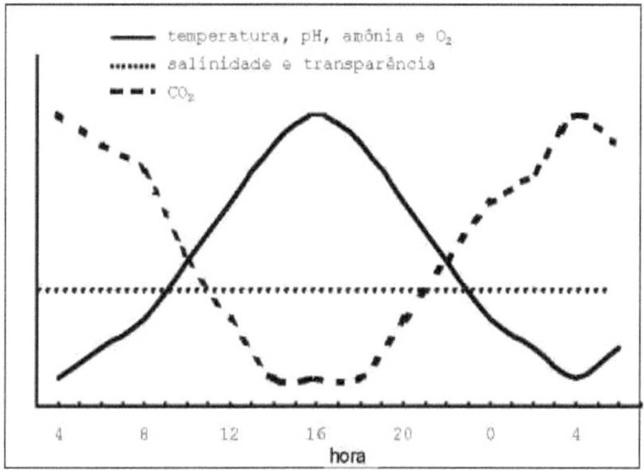

Figure 2 - Daily variations (arbitrary scale) in pH, dissolved O2, ammonia, temperature, CO2, salinity and transparency over a 24-hour period.

Source: KUBTZA, 2003; ARANA, 2004.

REFERENCES

ARANA, LV. Fundamentals of aquaculture. Florianópolis: UFSC Ed. 348p. 2004.

BOYD, CE; TUCKER, CS. Pond aquaculture water quality management.
Massachusetts: Kluwer Academic Publishing. 700p. 1998.

CAPUCCI, E. **et al.** Tubular wells and other groundwater abstractions. DRM-RJDepartamento de Recursos Minerais do Rio de Janeiro, 2001.

CREPALDI, DV; FARIA, PMC; TEIXEIRA, EA; RIBEIRO, LP; COSTA, AAP; DE MELO, DC; CINTRA, APR; PRADO, AS; COSTA, FAA; DRUMOND, ML; LOPES, VE; MORAES, VE. The situation of aquaculture and fishing in Brazil and worldwide. Rev. Bras. Reprod. Anim. 30(3/4), 81-85, 2006.

ESTEVES, FA. Fundamentals of limnology. Rio de Janeiro: Interciências (2ª Ed.) 575p. 1998.

KUBTIZA, F. Water quality in shrimp and fish farming. Jundiaí: CIP - USP Editora. 228p. 2003.

KUBITZA, F. Tilapia: production technology and planning. Jundiaí: Edcopyright, 2000. 289 p.

KUBTIZA, F. Water quality in fish farming. Jundiaí: CIP - USP Editora. 97p. 1999.

MARTINS, YK. Water quality in tilapia (**Oreochromis niloticus**) ponds: diurnal characterisation of physical, chemical and biological variables. Master's thesis from the Postgraduate Programme in Aquaculture and Fisheries at the Fisheries Institute - APTA - SAA. 54p. 2007.

PEREIRA, L; MERCADANTE, CTJ. Ammonia in fish farming systems and its effects on water quality. Boletim do Instituto de Pesca, São Paulo, 31(1), 81-86.

32

CHAPTER 4

BUILDING THE TANKS

Milton Erthal Junior

Alcimar Ribeiro Chagas

Rogério Burla Dias

Ana Paula Ribeiro Costa Erthal

4.1 CHOOSING THE SITE FOR THE TANKS

Within the property, it is recommended that the ponds be built on the highest points of the land in order to facilitate water management in the ponds, reduce the risk of flooding and minimise their exposure to the rising water table, which is shallow in the coastal lowlands of northern Rio de Janeiro. In addition, the water to be removed from the breeding ponds can be drained by gravity into ponds built in the lower parts of the land, where it must be stored and purified **before being disposed of in the soil or watercourses** or used in agriculture (ESTEVES, 1998).

It is advisable not to have trees or shrubs near the construction site of the tanks, as falling plant material in the tanks is harmful to the water, and some species with aggressive roots can puncture the plastic (KUBTIZA, 2000, 2003).

If the interested party wishes to build two or more tanks, it is recommended that a distance of around three metres be maintained between the tanks to facilitate unloading and circulation between the tanks in the area.

The location of the ponds must be well defined in order to facilitate the movement of cargo and the handling operations that are carried out on a daily basis, as well as facilitating the harvesting and transport of the fish.

4.2 BUILDING THE TANKS

The width of the tank depends on the width of the plastic to be used. In the trade, you can find reels 100 metres long and varying in width between 8 and 12 metres. The length of the tanks can be variable, adapting to the size of the land. Producers have been successful in raising tilapia in São João da Barra in tanks measuring 7 x 15 metres (width x length), which are easy to manage and can be built on plots, farms or small properties. To facilitate water flow, the bottom should be slightly sloping, with depths of 0.8 m and 1.0 m at the shallowest and deepest parts, respectively. In very shallow tanks, physical space for the fish is compromised, as well as being very susceptible to thermal oscillation. On the other hand, very deep tanks are more expensive to build and have a low water temperature at the bottom, which is an unfavourable condition for the fish, as well as not allowing higher fish densities. See the proposed dimensions in Figure 1.

During excavation of the land, the earth removed should be used to build the edge (dyke) of the tank, thus minimising the volume of earth excavated (Figure 1C). The edges should be levelled on the top surface, and before lining, roots, stumps and stones should be removed from the inside edges and bottom of the tank to prevent the plastic from tearing. A flat ruler should be used to smooth the bottom and edges, thus removing any ripples.

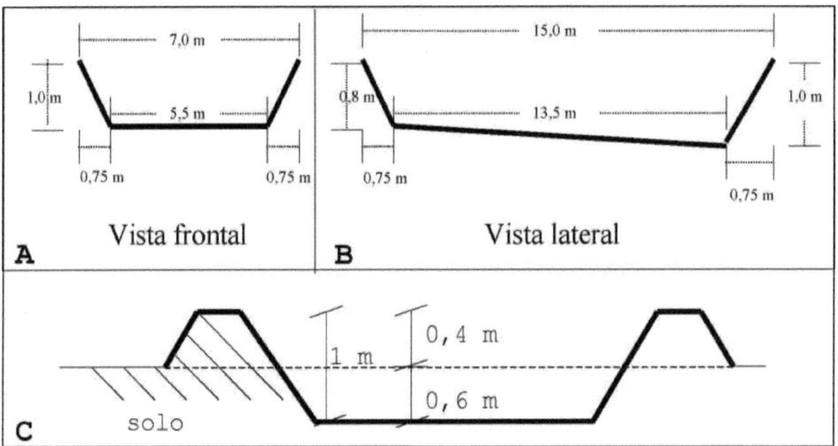

Figura 1.   Dimensions of the proposed tank. A - Front view; B - side view; C - front view showing the construction of the edge with the excavated earth.

At the edge of the deepest side of the pond, a slot should be dug for the installation of a 100 mm PVC pipe used in the drainage system (Figure 2). Once the horizontal pipe and the knee have been installed (Figure 2, C and D), the pond can be lined with plastic, which should be buried around the edges to secure it in place and grass planted above it to support the ground around the pond (Figure 3). It is recommended that the lining task be carried out in the early hours of the morning or on days with light winds to avoid wind disturbance.

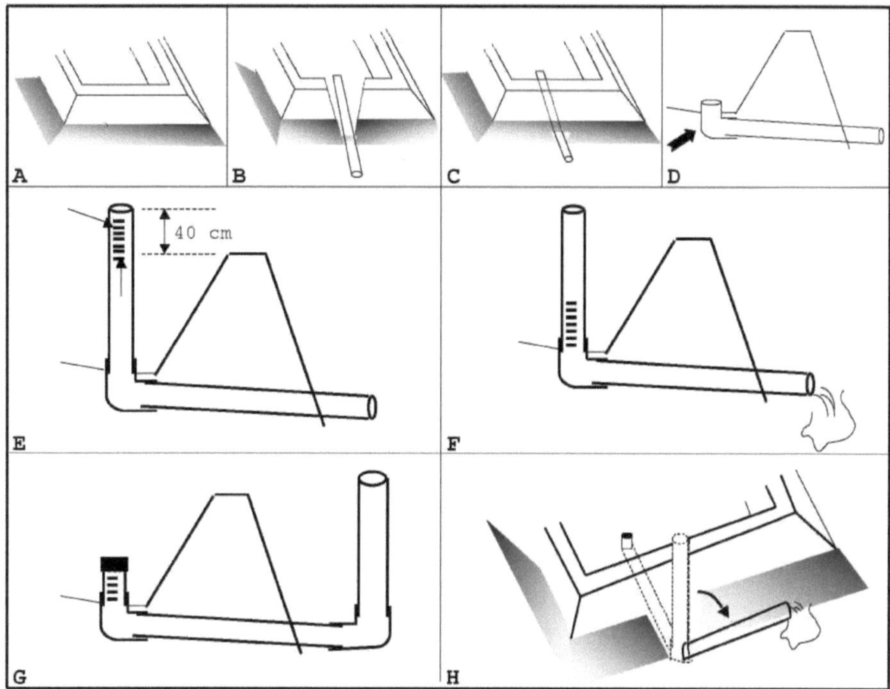

Figura 2. Suggestions for installing the tank's drainage system. **A** - view of the dam at the deepest part of the tank where the drainage system should be installed; B - opening a slit the dam to install a 100 mm pipe; C - reconstruction of the dam after installing the pipe; D - side view of the dam, showing the pipe together with the knee (arrow); **E** - installation of a 100 mm pipe over the knee with small slits (arrows) at the top to prevent the tank from overflowing due to carelessness; F - inverting the pipe to drain water from the tank; **G** - a second alternative for the drainage system, where a 30 cm pipe is placed over the knee, closed at the top and with slits for the water to escape. On the outside of the tank, a second knee and a pipe are installed, higher than the edge; **H** - to drain the water, simply turn the pipe, leaving it in a horizontal position.

After coating, a small hole (80 mm in diameter) should be drilled in the plastic just above the knee to install a 100 mm pipe 40 cm longer than the dyke (Figure 2 E). With the aid of a fine saw, commonly used for cutting metal, 10 cm long horizontal cuts should be made in the upper part of this pipe, i.e. the part that goes beyond the dyke, keeping 1 cm spacing between the cuts. Two or three rows of cuts should be made so that water can enter the pipe and allow it to drain. This prevents the tank from accidentally overflowing and the water can be drained by inverting the pipe (Figure 2 F). An alternative drainage system is proposed in tables G and H of Figure 2.

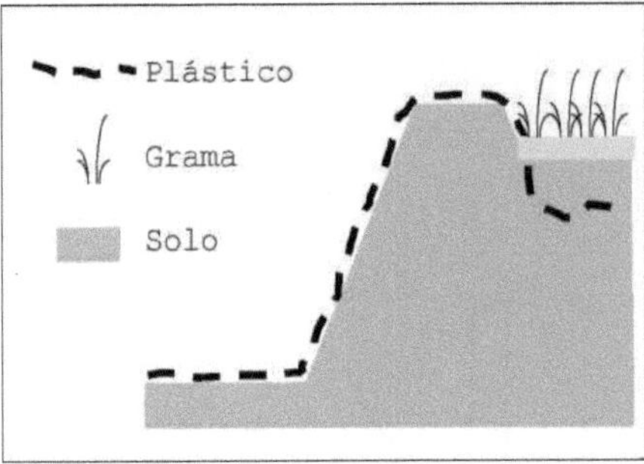

Figura 3.   Profile of the edge of the tank showing the plastic covered with soil with grass planted on top.

Then add water so that the plastic moulds to the tank. During the days when the tank is being filled, beware of animals (cats, dogs, horses, chickens, etc.) that could damage the plastic.

REFERENCES

ESTEVES, FA. Fundamentals of limnology. Rio de Janeiro: Interciências (2ª Ed.) 575p. 1998.

KUBTIZA, F. Water quality in shrimp and fish farming. Jundiaí: CIP - USP Editora. 228p. 2003.

KUBITZA, F. Tilapia: production technology and planning. Jundiaí: Edcopyright, 2000. 289 p.

KUBTIZA, F. Water quality in fish farming. Jundiaí: CIP - USP Editora. 97p. 1999.

CHAPTER 5

FERTILISING THE TANK

Milton Erthal Junior

Rogério Burla Dias

Ana Paula Ribeiro Costa Erthal

Fertilising the pond water should be done while the pond is being filled. This procedure aims to create suitable conditions for the development of phytoplankton and zooplankton, which are foods naturally used by tilapia. Growing tilapia in ponds with low water renewal, which is one of the characteristics of the system being proposed, has the advantage of reducing the cost of feed, since the natural food is of high nutritional value, containing carbohydrates, proteins, vitamins and minerals. Work from the 1990s showed that 30 to 70 % of fish growth can be achieved by consuming plankton and that when tilapia are grown in fertilised ponds, productivity increases by 2 to 12 times compared to unfertilised ponds (ESTEVES, 1998) (Table 1).The input used to fertilise the water can be of organic (animal or vegetable) or inorganic origin, and can be used together or separately. The most important issue in this process is that the final formulation has a C:N:P (carbon : nitrogen : phosphorus) ratio of 50:10:1 (KUBTIZA, 2003). As there is no product on the market with this constitution, it is necessary to combine two or more inputs.

Table 1. Productivity of some fish species when produced in ponds without and with fertiliser.

Species	Fish biomass (Kg/ha)	
	No fertiliser	With fertiliser
Tilapia	300 a 500	1000 a 3700
Common carp	250	1000 a 2000
Channel catfish	50	310 a 350
Pacu	-	300 a 800
Tambaqui	80	300 a 1600

Source: KUBITZA (2000)

Below are 9 different suggestions (letters A to I), all balanced, of organic and inorganic compounds that can be used to fertilise 15 x 7 m ponds.

A. 2.5kg of cattle manure + 0.415g of urea.

B.   1.25 kg of pure poultry manure + 0.420 kg of urea.

C.   2 kg of pig manure + 0.380 kg of urea.

D.   1kg of earthworm humus + 0.5kg of urea.

E.   300 g of simple superphosphate + 550 g of urea.

F.   420 g of N-P-K (4-14-8) + 500 g of urea.

G.   210 g of N-P-K (8-28-16) + 520 g of urea.

H.   170 g of N-P-K (10-34-00) + 520 g of urea.

I.   160 g of N-P-K (11-37-00) + 520 g of urea.

Don't mix formulations to make a single fertiliser. Different formulations can be used throughout the growing season.

Fertilising the ponds should be repeated every 15 days if the greenish colour of the water is not maintained. Afterwards, the fish faeces themselves will serve as a substrate for phytoplankton and zooplankton.

Of the options recommended above, it is suggested that the producer use the one that is easiest to acquire or least expensive. However, the first fertilisation should be done with an organic fertiliser and supplemented with urea (items A, B, C or D) to provide the system with carbon, since there is no water contact with the soil as in traditional crops. An additional factor is the presence of micronutrients and minerals in the manure, which in a single application can meet demand throughout the crop cycle.

<u>Important note</u>

It is prudent to monitor ammonia and dissolved oxygen in the water in the days after fertilisation to avoid surprises, such as fish kills.

REFERENCES

ESTEVES, FA. Fundamentals of limnology. Rio de Janeiro: Interciências (2ª Ed.) 575p. 1998.

KUBTIZA, F. Water quality in shrimp and fish farming. Jundiaí: CIP - USP Editora. 228p. 2003.

KUBITZA, F. Tilapia: production technology and planning. Jundiaí: Edcopyright, 2000. 289 p.

KUBTIZA, F. Water quality in fish farming. Jundiaí: CIP - USP Editora. 97p. 1999.

MARTINS, YK. Water quality in tilapia (**Oreochromis niloticus**) ponds: diurnal characterisation of physical, chemical and biological variables. Master's thesis from the Postgraduate Programme in Aquaculture and Fisheries at the Fisheries Institute - APTA - SAA. 54p. 2007.

CHAPTER 6

TILAPIA PRODUCTION PLANNING

Milton Erthal Junior

Alcimar Ribeiro Chagas

Ana Paula Ribeiro Costa Erthal

6.1 Production in stages

Tilapia production in phases optimises productivity per area and can increase it by up to 38% (KUBTIZA, 2000). The basic principle of phased production is to ensure that the biomass of fish per pond does not exceed the CRITICAL BIOMASS[1] . In the 3-phase system, it is possible to grow three or more cycles per year, while in the single-phase system, the producer can grow a maximum of two cycles per year. This is possible by acquiring a new batch of fry as soon as the fish are transferred to the finishing pond (105 m^2), so that all three ponds are always occupied with fish.

6.2 Densities

The densities of fish in each phase are described in Figure 1. If the system used is a single-stage system, it is advisable to use 1 to 2 fish/m^2. The densities proposed in this work were suggested according to (KUBTIZA, 2000). In order to obtain the highest density per tank, good management and monitoring of water quality are necessary to avoid under-growth, size unevenness or even sudden mortalities.

As mentioned in CHAPTER 3 (Water Quality), the most limiting factors in this system are the concentration of O_2 and ammonia in the water, which determine the density in the tank. These problems can be overcome by installing emergency aeration systems and/or more frequent water renewal. These measures result in increased productivity and production costs. In this way, the method of commercialisation can be a determining factor in whether the producer takes the risk of investing for greater productivity.

---

[1] CRITICAL BIOMASS: this is when the daily growth of the fish reaches its maximum value, i.e. maximum of daily weight gain per unit area (Kg/m^2/day). From this point onwards, growth is increasingly reduced until the pond's SUPPORT CAPACITY is reached, at which point the fish no longer grow (KUBTIZA, 2000).

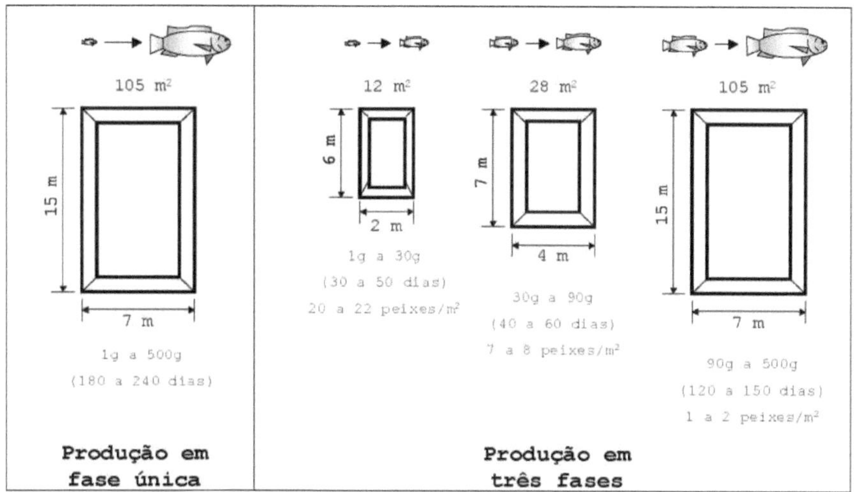

Figure 1. Single-stage and three-stage tilapia farming. The proposed dimensions, expected growth, time in each phase and density were recommended for cultivation systems with little water renewal and the addition of complete feed, considering a final density of 2 fish /m².

Harnessing wind energy through the use of wind turbines is an interesting alternative for pumping water into the pond at low cost, allowing for increased productivity. On rural properties where irrigation is used for agriculture, it is recommended that water be collected from the bottom of the pond, thus removing water that is bad for the fish but of excellent quality for the plants. If the water is used frequently, so that the water in the pond is frequently renewed, the densities of the crop can be increased.

6.3 Purchase and transport of fingerlings

Fry should be purchased from reputable producers with a proven track record in the market. This is important, because buying poorly sexed fry can lead to serious fattening problems due to disordered breeding in the pond. A batch of well-reversed fry is accepted when more than 98 per cent of the population is male. The advantage of using males for fattening is that they grow 1.8 to 2.5 times faster than females (KUBTIZA, 2003). In addition, during reproduction, females channel their reserves into egg production, eat little during egg incubation, which takes place in the mouth (2 to 3 weeks) and overcrowd the tanks, which results in competition for food, drastically reducing the development of the fish.

For transport, the fry should be packed in containers with water and oxygen (plastic bags are suitable for this purpose). During the journey, it is recommended that the fish are protected from the sun. Before introducing them into the tank, keep the bags in the water for about 20 minutes to balance the temperature and avoid thermal shock. Then gradually add the water from the pond into the bag containing the fry before releasing them.

More details on this procedure are given in Chapter 2 of this book.

REFERENCES

KUBTIZA, F. Water quality in shrimp and fish farming. Jundiaí: CIP - USP Editora. 228p. 2003.

KUBITZA, F. Tilapia: production technology and planning. Jundiaí: Edcopyright, 2000. 289 p.

KUBTIZA, F. Water quality in fish farming. Jundiaí: CIP - USP Editora. 97p. 1999.

CHAPTER 7

FEEDING TILAPIA

Milton Erthal Junior

Alcimar Ribeiro Chagas

Ana Paula Ribeiro Costa Erthal

Tilapia are omnivorous, opportunistic fish with a tendency to herbivory (STARLING, 1998), i.e. they efficiently use the algae present in the aquatic environment to satisfy their nutritional needs. In intensive tilapia farming systems, primary production (phytoplankton) is not enough to nourish the fish, making it necessary to use feed. Unlike many fish species, tilapia accept feed in different forms (pelleted, pelleted and extruded) and have the ability to produce high-quality proteins from alternative, lower-cost sources that give satisfactory results in fish performance.

Normally, the fry acquired are on average 1.5 to 2.0 cm long, weighing less than 0.5 g each. During this phase, up to approximately 5 to 6 cm, the fish should be fed a mealy ration containing 35% crude protein, adding 5% of the fish's biomass daily (KUBTIZA, 2000).

For example: 500 fish averaging 2.0 cm, each weighing 0.5 g = 250 g of total biomass; 5% of the biomass = 12.5 g of pelleted feed/day. As these fish still have a small alimentary canal, it is advisable that this quantity (12.5g/day) be divided up so that it is offered several times a day. For example, five times a day: 9am = 2.5g; 10am = 2.5g; 3pm = 2.5g; 4pm = 2.5g and 6pm = 2.5g.

It is advisable to avoid feeding them between 10am and 3pm, as the sun's rays can affect the fish's skin and mucous membranes, causing injury (KUBTIZA, 2000).

As there is only a limited supply of meal in the shops, we advise you to buy extruded starter feed (35 per cent protein) and grind it in a blender.

From 5 to 6 cm in length, the fish will be fed the same initial extruded feed (35% protein), but in pellet form. At this stage, until the fish reach 13 to 15 cm, this feed should be added at a rate of 5 % of their total biomass, divided into 4 daily schedules.

To find out the biomass, an aliquot of fish (5 to 10 % of the pond's total) should be weighed every 15 days to estimate the pond's biomass and monitor the fish's development. If you don't have a scale, the fish can be measured at their total length (mouth to the end of the trough) (Figure 1) to estimate their weight according to Table 1.

Fish ranging in size from 15 cm (50 g) to 24 cm (300 g) will be fed an intermediate ration consisting of 28 % protein, 3 times a day, adding 2 % of the total biomass.

As soon as the fish reach 300 g (24 cm), they will be given a finishing ration containing 22 per cent protein, 2 per cent of the total biomass, divided into 4 schedules.

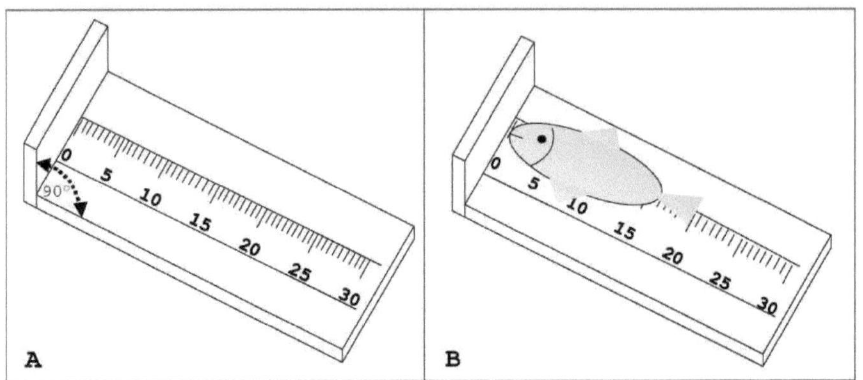

Figure 1 - Ichthyometer used to monitor fish length during rearing. It is an easily constructed tool, requiring two pieces of wood, which must be nailed together to form a 90° angle between the pieces, and an ordinary ruler, which must be adhered to the top surface of the longer piece of wood (A). Illustration of the length of a tilapia, with 20 cm between the mouth and the tail (B).

Table 1. Estimated relationship between mean length, weight and age of well-fed hybrid tilapia.

Total length (cm)	Weight (g)	Age (weeks)
0,5	0,005	0
1	0,003	1
2	0,17	2
3	0,31	3
4	1,0	4
5	2,0	5
6	5,0	6
8	10,0	
10	17,0	
12	26,0	14
13	33,0	
15	50,0	18
16	68,0	
17	90,0	
18	118,0	22
20	165,0	
22	225,0	
24	300,0	
25	345,0	
27	440,0	29
29	490,0	> 29

Source: Yance & Menezes, 1982

<u>IMPORTANT NOTE</u>

The amount of feed suggested above for the phases is a theoretical estimate and can vary greatly between ponds or farms. Considering that feed represents around 60 per cent of the cost of tilapia production, any savings on this item can represent an increase in the profit margin for the producer. In this way, a pond with a large volume of phytoplankton (green algae) can save on feed. Therefore, the feeder (person who feeds the fish) should observe whether there is a surplus or an undersupply of feed in the ponds so that it is offered in the right measure in each pond. In the case of tanks with phytoplankton, 90 per cent of the amount required by the fish should be offered to stimulate algae consumption. To find out how much each pond consumes, a table should be drawn up showing the amount of feed added to each pond each day at the four feeding times.

The form of feed can vary depending on the region, as the climate is different, the variety of tilapia and the quality of the feed. It is therefore advisable to carry out a regional study of tilapia production by specialised technicians and/or rural extension agencies before starting farming, in order to increase the reliability of the procedure.

7.1 Zootechnical indices to be monitored

Monitoring fish development can reveal whether management is being carried out satisfactorily. The most common parameters for monitoring production are: average daily gain (ADG), feed conversion (FC) and survival rate (SR).

$$\text{GMD} = \frac{\text{Final weight (g) - Initial weight (g)}}{\text{Cultivation period}}$$

$$\text{CA} = \frac{\text{Food consumed (Kg)}}{\text{Weight gain (Kg)}}$$

$$\text{TS} = \frac{\text{At the end of fish} \times 100}{\text{Initial N}^{\circ}\text{ of fish}}$$

Bibliographical references

ASSUMPÇÃO, J. & NASCIMENTO, M.T. Structure and floristic composition of four restinga vegetation formations in the Grussaí/Iquipari lagoon complex, São João da Barra, RJ, Brazil. **Acta bot. bras.** 14(3): 301-315. 2000.

CAPUCCI, B.E.; MARTINS, M.A.; MANSUR, L.K., MONSORES, M.L.A. **Tubular wells and other groundwater abstractions** - Guidance for Users. PLANÁGUA Project SEMADS/GTZ. Niterói, Department of Mineral Resources of the Government of the State of Rio de Janeiro, 2001. 70p.

FREITAS, R.T.F.; LAGO, A.A.; ALLAMAN, I.B. & REIS NETO, R.V. **Exotic and native species of importance to Brazilian fish farming.** Minas Gerais: UFLA: FAEPE, 2009. 65p.

KUBITZA, F. Tilapia: technology and planning in commercial production. Jundiaí: F. Kubitza, 2000. 285p

STARLING, F.L.R.M. 1998. **Development of biomanipulation strategies for remediation of eutrophic problems in urban reservoir, Lago Paranoá, Brazil.** Doctoral thesis. Institute of Aquaculture. University of Stirling, 226 p.

YANCE, D.R. & MENEZES, J.R.R. **Manual de criação de peixes.** Fundação Cargil, 1ª Edição, Campinas, 110 p. 1982.

CHAPTER 8

AQUAPONY

Luiz Paulo de Oliveira Tostes

Felipe Cabral da Costa de Barros

Milton Erthal Junior

Ana Paula Ribeiro Costa Erthal

8.1 INTRODUCTION

Aquaculture is a multiple activity that refers to the cultivation of various aquatic organisms: aquatic plants, molluscs, crustaceans and fish. In this breeding process, it is trivial to intervene or manage in order to increase production as a result (OLIVEIRA, 2009). If their effluents are discharged into **watercourses** without prior treatment, **this** activity has the potential to pollute natural resources (ASSUNÇÂO, 2011). Therefore, in order to move towards sustainable aquaculture production, the industry must recognise and address all the environmental impacts caused by its operations (Ministério da Pesca e Aquicultura, 2012).

The principle of aquaponics is to make use of leftover food and fish waste, which are transformed into nutrients for the plants, which in turn filter the water before it is returned to the breeding ponds (RAKOCY and HARGREAVES, 1993). Even though it is not very common in Brazil, it has great potential to be implemented, since the increase in the number of commercial fish farmers, the high cost of fertilisers used in hydroponics and the irrational use of water in agriculture have become great allies in defending this activity as a sustainable and profitable source (TESTOLIN, 2009).

8.2 AQUAPONY

The intensification of the cultivation of some fish species in a super-intensive production system, such as tilapia, is due to the reduction in the capture of marine species and because tilapia farming has some advantages, such as prolificacy, hardiness and weight gain. This system does present risks, such as

the amount of food not used by the fish and faecal solids in the form of organic waste. Discharging these effluents into nature causes serious environmental damage such as: pollution, hybridisation, the appearance of exotic species, excesses of algae, deterioration of water quality and soils (CRIVELENTI, BORIN and SILVA, 2009). However, this waste can be used to nourish hydroponic plants, thereby purifying the water and making it fit to return to the fish ponds (NOGUEIRA FILHO, et al 2003).

Before it was known as aquaponics, the Aztec Indians had difficulty finding a place to grow their food. So they built rafts that floated across Lake Tenochtitlan, which were covered with earth dredged from the bottom of the lake, and from there they grew their vegetables. As the plants grew, their roots came into contact with the water (NELSON AND PADE, 2008).

Increasingly popular, aquaponics is already in the catalogues of major suppliers of aquaculture and hydroponics materials, has been published in a quarterly journal, and some schools already include aquaponics in their science lessons (RAKOCY, et al 2006).

According to Rinehart (apud Diver, 2010) aquaponics is a system that integrates fish farming with the hydroponic cultivation of vegetables through the recirculation of water in a closed environment, serving as a model for the sustainable development of production and providing healthy food, as well as improving the local economy. Systems that involve two distinct productions are considered complex, as they combine antagonistic systems. The biomass of the fish is aquatic and the phytomass of the plants is **non-aquatic**, so strict control of **the system** is necessary **for the good growth of both species. "Aquaponics** is the production of fish associated with the production of vegetables. With the current global trend towards preserving the environment, and especially water resources, this is an activity that, in addition to reducing the use of inputs, utilises the reuse of water in a **production system". (TESTOLIN, 2009),**

CORTEZ (1999) mentions studies in intensive fish farming that use leftover feed and fish excrement dissolved in the water to grow hydroponic vegetables, improving the quality of the water before it is returned to the fish rearing reservoir.

Uneaten food or any organic nitrogenous waste in the system is transformed into ammonia by the phenomenon of ammonisation by decomposer bacteria. Fish excrement releases nitrogen into the system, and bacteria of the Nitrosomanos type carry out the Nitrosation phenomenon and convert ammonia into nitrite, which is still toxic to fish. Subsequently, the Nitrobacter bacteria carry out another phenomenon, nitration, transforming nitrite into nitrate, which is relatively harmless and is the form in which plants absorb nitrogen for their development (ABREU, 2011).

Therefore, the excess nutrients produced by fish farming, which is presented as a pollutant, is a rich source of nutrients to serve as food for plants, which in turn have the important function of filtering the water. There are still discussions about whether this cycle is self-sufficient (ABREU, 2011). In the majority of experiments on integrated systems, deficiencies in some nutrients have been noted, (RAKOCY , 1993) who pointed to low iron concentrations, CORTEZ et al (2008) found deficiencies in potassium and magnesium. However, SANTOS et al (2007) obtained better lettuce yields by adding 25% hydroponic solution to their experiment, generating great savings in the purchase of nutrient solutions without jeopardising fish cultivation. Although there are discussions about the development of plants in aquaponics, the advantages of integrating these two cultures are clear and evident (ABREU, 2011).

Still not widely known in Brazil, aquaponics has great potential to be implemented in the country due to the growth of commercial fish farmers, the expensive industrial fertilisers used to nourish hydroponic systems, the excessive use of water in agriculture, the fact that it does not harm the environment (TESTOLIN, 2009) and also offers a complete meal with fish protein, fibre, nutrients and minerals from vegetables, fruit and herbs (CONNOLLY AND TREBIC, 2010).

REFERENCES

ABCC. Panorama of World and Brazilian Fish Production, with Emphasis on the Aquaculture Segment. Available at:<www.abccam.com.br/ >. Accessed on: 16 Apr. 2012.

ABREU, S.P.M. Technical dossier: aquaponic system, Centre for the support of technological development CDT/Unb - SBRT - 2011

ARANA, L. V.; Chemical principles of water quality in aquaculture: a review for fish and shrimp. Florianópolis: UFSC, 1997. 166 p.

AZZOLINI, J. C.; ZARDO, F; SEGALIN, C. A.; Monitoring of effluents from fish pond discharge at different levels of discharge. Unoesc & Ciência - Acet, Joaçaba, v. 1, n. 1, p.59-68, 2010.

NATIONAL DEVELOPMENT BANK. Panorama of aquaculture in Brazil: challenges and opportunities: Agroindustry. BNDES Sectoral 35, p. 421 - 463. Available at: <www.bndes.gov.br/ >. Accessed on: 01 April 2012.

BARROS, F. G. N.; AMIN, M. M.; Water: scarcity, valuation and prospects for brazil. Available at: <http://www.sober.org.br/palestra/2/162.pdf>. Accessed on: 26 March 2012

CAMPANILI, M.; In Brazil, there is a deficit in the midst of abundance. São Paulo: Agência Estado, Caderno Ciência, 2003. Available at http://www.estadao.com.br/ext/ciencia/agua/aguanobrasil. Accessed on 23 Feb 2005.

CONNOLLY, K; TREBIC, T.; Optimisation of a backyard aquaponic food production system. Faculty Of Agricultural And Environmental Sciences Macdonald Campus, Mcgill University: Bree 495 Design 3 Bioresource Engineering Presented To Dr Vijaya Raghavan, 2010. 74 p.

CORTEZ, G.E.P.; Growing lettuce in hydroponics associated with fish farming. 1999. Jaboticabal, 63f. Thesis (Doctorate in Agronomy) - Postgraduate Programme in Agronomy, São Paulo State University - UNESP - Jaboticabal campus.

CRIVELENTI, L.Z.; BORIN, S.; SILVA, N.R. Superintensive fish farming associated with hydroponics in a recirculating water system. Archives Of Veterinary Science,: Nota Científica, Uberlândia, v. 14, n. 2, p.109-116, 19 jun. 2009.

DIVER, S.; Aquaponics - integration of hydroponics with aquaculture. ATTRA - National

Sustainable Agriculture Information Service, Fayetteville, AR, USA. 28 pp. 2010

DUARTE, E.; Cultivation of nile tilapia post-larvae using different proportions of shell/gravel substrate in the biofilter. 2010. 47 f. Dissertation (Master's Degree) - Animal Science Course, Universidade Federal Dos Vales do Jequitinhonha e Mucuri, Diamantina, 2010.

GIA. Sector study to consolidate sustainable aquaculture in Brazil: 2007. Available at at: <ftp://ftp.fao.org/fi/document/aquaculture/sect_study_brazil.pdf>. Accessed on: 23 Apr. 2012.

MASUDA, C. T.; Trends and prospects for fish production in Brazil. 2009. 55 f. Monograph (Bachelor's Degree) - Veterinary Medicine Course, Faculdade Metropolitanas Unidas, São Paulo, 2009

MPA. Fisheries and aquaculture statistical bulletin: Brazil 2010. Available at: <www.mpa.gov.br/ >. Accessed on: 29 January 2012.

NELSON, R. L.; PADE, J. S. Aquaponic food production. Montillo, Wi. Usa. 2008. 218 p

NOGUEIRA FILHO, Hercules et al. Aquaponics: interaction between hydroponic lettuce and super-intensive Tilapia farming.Available at: <www.abhorticultura.com.br/biblioteca/arquivos/.../cpfg2025c.pdf >. Accessed on: 12 Feb. 2012.

OLIVEIRA, R. C.; O panorama da aquicultura no brasil: a prática com foco na sustentabilidade. Revista Intertox de Toxicologia, Risco Ambiental e Sociedade, São Paulo, v. 2, n. 1, p.71-89, 01 feb. 2009.

RAKOCY, J. E., MASSER, M. P., & LOSORDO, T. M. Recirculating Aquaculture Tank Production Systems: Aquaponics-Integrating Fish and Plant Culture. Southern Regional Aquaculture Centre, Publication No. 454, 1-16. 2006

RAKOCY, J., R.C. SHULTZ, D.S. BAILEY, E.S. AND THOMAN. Aquaponic Production of tilapia and basil: comparing a batch and staggered cropping system. Acta Horticulturae. Vol. 648. p. 63-69, 2006

RAKOCY, J.E.; HARGREAVES, J.A. Integration of vegetable hydroponics with fish culture: a review. Techniques for Modern Aquaculture, SRAC Publication No. 453 p.112-136,1993

REBOUÇAS, B. B., CUNHA. A., BRAGA B. Águas doces no Brasil: capital ecológico, uso e conservação. São Paulo. 1999.

SEBRAE. Santa Catarina in numbers: fishing and aquaculture - 2010. Available at: <www.sebrae.com.br>. Accessed on: 23 March 2012.

SHIMODA, E.; ANDRADE, D.R.; Fish farming: water quality monitoring and control practices. Technical Bulletin - UENF. V 3. N 3 Campos dos Goytacazes. 1998.

TESTOLIN, G.; Evaluation of hydroponic lettuce using fish farm water mixed with different percentages of nutrient solutions. 2009. Piracicaba, 76f. Dissertation (Master's Degree in Agronomy) - University of São Paulo - **"Luiz Queiroz"** School of Agriculture.

UNIAGUA. Water on the planet. Available at: < http://www.uniagua.org.br/website/default.asp?tp=3&pag=aguaplaneta.htm>. Accessed on: 06/04/2012

VALENTI, W. C. 2002. Sustainable aquaculture. In: Zootechnics Congress, 12th, Vila Real, Portugal, 2002, Vila Real: Portuguese Association of Zootechnical Engineers. Proceedings...p.111-118.

CHAPTER 9

WATER AVAILABILITY IN THE LOWER PARAÍBA DO SUL

Vicente Paulo dos Santos

9.1 Water in the World

Water is a limited renewable natural resource. It is known that in all civilised regions water is increasingly a rare and therefore expensive commodity. Estimates made by the International Hydrological Programme (UNESCO/PHI, 1998) indicate that of the approximately 1.4 billion cubic kilometres of water on Earth, 97.5% is salt water and only 2.5% is fresh water, which is essential for humanity. This means that the amount of fresh water on Earth is around 34.6 million cubic kilometres. Of this volume, 68.9% or around 24 million cubic kilometres form the polar ice caps and glaciers. However, these springs occur in solid form and are located at great distances from the Earth's inhabited areas. Groundwater, on the other hand, represents around 29.9% of the Earth's total freshwater volume (around 10 million cubic kilometres) and occurs in liquid form. The volume of fresh water available in rivers and lakes is only 0.3% of this total, or around 104,000 cubic kilometres.

As it is a finite resource, there is no possibility of unlimited consumption. Therefore, if consumption patterns are not reduced, it will not be possible to guarantee its availability indefinitely.

According to Assis (2000), South America has around 47 per cent of the planet's water resources (fresh water), mainly concentrated in Brazil, **which has around 100,000 watercourses and yet treats them with** contempt - even though studies indicate that water is set to become the planet's scarcest and most precious commodity in the 21st century.

Anticipating a situation that is likely to worsen in the future, the world's water supply no longer meets the demand for all multiple uses, with around 25 per cent of the population in developing countries having no access to drinking water and sewage systems.

The gravitational force causes the water in the atmosphere to precipitate, run off and infiltrate through the surface of the ground, remaining stored underground until it is drained **by the watercourses that form the river basins. The portion of water that** infiltrates and moves through the subsoil is important for feeding the basic flow of **watercourses. However, the irregularity of its distribution -** both in space and time - in relation to the needs or demands of the populations in the respective areas, can further reduce its availability.

Appropriate land use, favouring infiltration, makes it possible to maintain a greater volume of stored water, which allows for greater control of the **fluctuations in flows in watercourses, as it**

reduces the maximum flows produced in them and makes it possible to store a greater volume of water to be made available in dry periods.

In agricultural production systems, the problem is associated with keeping precipitated water on the property. Much of this water flows over the surface of the soil, reducing the volume of water that reaches the water table. De Maria (1999) estimated water losses in areas with agricultural crops at 2,519 $m^3.ha^{-1}.year^{-1}$ and, for areas under pasture, equivalent to one tenth of this, corresponding to a loss of around 171 billion m^3 of water per year in areas occupied by these types of uses. This water loss reduces the volume of water available for plants, as well as that used to supply rivers and wells.

Water erosion promotes the transport of sediment and chemical or organic fertilisers into water bodies, causing them to become polluted and contaminated, increasing water treatment costs, reducing the potential for electricity generation and consequently raising the cost of these products. Contamination of water bodies also has a direct impact on the health of users of contaminated water, as well as reducing aquatic life and causing environmental imbalances. The high rate of surface runoff and the low capacity of natural or artificial drains also causes flooding in rainy periods and water shortages in dry periods.

The use of water is becoming increasingly problematic, also due to the precariousness of basic sanitation systems, the insufficient quantity and high costs of the treatment systems needed to meet drinking water standards.

9.2 Water in the Lower Paraíba do Sul

In 1990, the United Nations (UN) identified around twenty countries with water scarcity problems and predicted that by 2010 fifteen more would suffer from the same problem if the current situation of degradation continued and if measures were not taken to preserve and rationalise the use of water resources.

In our country, the planning and management of water resources is becoming increasingly important as economic development and the resulting increase in demand for water generate conflicts over its use.

Law no. 9.433 of 8 January 1997, which instituted the National Water Resources Policy, has as its main foundation the recognition of water as a finite, vulnerable resource, endowed with economic value, with the aim of guaranteeing ecological, administrative and financial sustainability, restating Brazilian society's commitment to water bodies. In the state of Rio de Janeiro, basin committees were created to make up the State Water Resources Management System, established by State Law No. 3.239/1999, whose objectives are: to settle, in the first instance, any conflicts relating to the use of water; to monitor the State Water Resources Policy; to propose values and approve criteria for charging for the use of water; and to plan, regulate and control the use, preservation and recovery of water resources.

In Baixada Campista, located in the Lower Paraíba region, Hydrographic Region IX of the state of Rio de Janeiro, belonging to the Paraíba do Sul River Basin, water problems relate not only to its availability but also to its distribution. The region is well served in terms of water resources, including the Paraíba do Sul River, the Muriaé River, the Ururaí River, and Lagoas Feia, de Cima and Campelo, among others. The system also has other smaller rivers, as well as an extensive network of canals, some 1,500 kilometres long, implemented by the now defunct National Department of Works and Sanitation (DNOS), which serves a vast lowland region with great agricultural potential.

Due to unsustainable practices such as deforestation, especially with the removal of riparian forests, sugar cane monoculture, fires and others, combined with low investment in agricultural technology, these natural resources have reduced the availability of water. The difficult maintenance of the network of canals set up by DNOS has meant that the agricultural sector in the Baixada Campista has suffered major losses, especially during periods of drought, because the lack of maintenance of the system has meant that this structure cannot be used for agricultural production.

9.3 The Paraíba do Sul River Basin

The Paraíba do Sul river basin covers an area of 62,074 km^2, between the states of São Paulo, Rio de Janeiro and Minas Gerais. The main channel of the river is formed in the state of São Paulo and runs through the state of Rio de Janeiro, delimiting its border with the state of Minas Gerais along the mountainous region. As such, the Rio de Janeiro portion of the Paraíba do Sul river basin is characterised by being downstream of the São Paulo portion, formed mainly by the Paraitinga and Paraibuna tributary rivers, and the Minas Gerais portion, formed mainly by the Preto, Paraibuna, Pomba and Muriaé tributary rivers. The Paraíba river basin, sub-basin 58, is part of the Eastern Atlantic Basin according to ANEEL's hydrological classification. The river has an average flow at the city of Campos of 876 m^3/s, relating to a period from 1934 to 1992 (ANEEL, 2000), and minimum and maximum flows of 115 and 6,000 m^3/s, respectively, considering a return period of 50 years (PROJIR, 1982).

With regard to the quality of its waters, previous studies indicate that they are classified, according to Salinity Laboratory criteria, as C1S1. The Paraíba do Sul river and its tributaries, which originate in the state of Minas Gerais, reach the state of Rio de Janeiro with water quality already compromised by the discharge of domestic sewage, industrial effluents and a high load of suspended solids. In the Rio de Janeiro stretch of the basin, sanitary sewage contributions that reach the water bodies **in natura** and industrial discharges add to the pollutant load, worsening water quality. Despite the control that the State Environment Institute (INEA) exercises over the more than 8,500 industries in the basin, the effects of toxic substances and heavy metals are still significant in the waters of the

Paraíba do Sul river. The problem of rubbish dumps, deforestation and consequent erosion, the removal of mineral resources for civil construction without proper environmental recovery, accidents involving the transport of toxic cargo, the devastation of riparian forest, predatory fishing, the improper and uncontrolled use of pesticides and a lack of environmental awareness are also among the factors contributing to the degradation of environmental quality in the basin, particularly its water resources.

With regard to basic sanitation, according to CEIVAP (the Paraíba do Sul River Basin Integration Committee), the situation of degradation is critical, as currently around 1 billion litres of domestic sewage, practically untreated, are discharged daily into the rivers of the Paraíba basin - 90% of the municipalities in the basin do not have a sewage treatment plant. In addition to domestic effluents, there are 150 tonnes of BOD (Bio-Chemical Oxygen Demand) per day, corresponding to the pollution load derived from organic industrial effluents (not counting toxic agents, mainly heavy metals). The Paraíba basin's total pollution load of organic origin corresponds to around 300 tonnes of BOD per day, of which around 86% comes from domestic effluents and 14% from industrial effluents. Despite the compromised water quality, approximately 14.2 million people, plus the 8.7 million inhabitants of the metropolitan region of Rio de Janeiro, are supplied by the waters of the Paraíba do Sul River Basin.

The Lower Paraíba do Sul region, near the mouth of the basin, is characterised by the presence of lagoons and canals. On the right bank of the Paraíba do Sul river, in the stretch corresponding to the urban area of the city of Campos dos Goytacazes up to the mouth in the municipality of São João da Barra, there are currently around 1,500 kilometres of canals that are responsible for draining the city and the Baixada Campista, some of which also function as irrigation canals, taking water from the Paraíba river for this purpose. The existing network of canals is the result of sanitation projects, implemented from 1933 onwards by the then Baixada Fluminense Sanitation Board, a body replaced by the now defunct DNOS (National Department of Sanitation Works), with the aim of controlling the flooding of the Paraíba do Sul River and expanding the region's arable areas, drastically reducing the natural wetlands. By draining swamps and marshes, opening and enlarging water drainage channels, directing the river's overflow into Lagoa Feia, and artificially opening the bars of lagoons and lagoons, these projects led to the destruction of entire aquatic ecosystems, the reduction of the surface area of lagoons, lagoons and floodplains.

The left bank of the Paraíba do Sul river in the stretch corresponding to the urban area of the city of Campos dos Goytacazes has several channels and ditches interconnected by the lagoon system that exists there (Lagoas do Vigário, dos Prazeres, Brejo Grande and Campelo), thus forming the local drainage network.

In addition to the Paraíba do Sul River, the region also has other water resources, including the Cima Lagoon, the Feia Lagoon and a series of rivers, which make it a potential agricultural producer. However, the environmental degradation caused by a series of factors such as sugar cane monoculture, inappropriate agricultural practices, deforestation and disorderly urban growth have left the region's soils unprotected, reducing their storage capacity, which can be seen during periods of drought.

The canals currently have a series of problems, which clearly reflect the complexity of the system. The problems start close to the catchment area, in the urban area, where silting, the presence of rubbish, the reception by the canals of various effluents, mainly sanitary sewage, make their waters unsuitable for agricultural use, as well as reducing the availability and often preventing the transport of water. In rural areas, the problems that occur are actually the result of improper use in urban areas, i.e. a lack of water, which causes a drop in agricultural production, salinisation of lagoons and ponds, and consequently fish kills, causing a picture of desolation in a region surrounded by water of sufficient and necessary quantity and quality.

As for the canals, the primary ones are Campos-Macaé, Coqueiros, Cambaíba, São Bento and Quintinguta and the secondary ones are Tócos and Nicolau.

Of these channels, a special situation is observed in the Quintinguta area. This canal passes entirely through the restinga area of the complex, and is subject to the saline tongue of the sea, which invades via the water table and direct entry, mainly at Barra do Açu. With drought, it is possible to see salinisation of the environment in these flooded areas, resulting in fish deaths and a drop in agricultural production. Next to this canal, in the municipality of São João da Barra, the Port of Açu Logistics and Industrial Complex is being installed, a major undertaking that will demand water from the region's existing water resources.

9.4 Channels of the Baixada Campista

The Baixada Campista has extensive plains with high fertility, favourable for growing sugar cane, fruit and livestock. The Paraíba do Sul River, in its sedimentary stage, flowed through the municipality at a higher level than in the Baixada region, favouring the occurrence of floods and other sanitation problems in the past. To solve these problems, the National Department of Works and Sanitation - DNOS, which was abolished in 1989, built an extensive network of artificial canals with the aim of draining and sanitising the flooded areas of the Baixada Campista. The implementation of the canal network led to the expansion of the region's agricultural frontiers, and the system was also used for irrigation purposes in times of drought, mainly from the Paraíba do Sul river.

With the extinction of DNOS, the maintenance and control of the network was compromised due to abandonment. As a result, its use became chaotic, with occasional interventions according to private interests, causing inefficiency in the function of the canals, as well as a series of conflicts over the use of water.

In 1997, as a result of the major flooding that occurred in the region, Campos City Hall, through the Department of the Environment and Civil Defence, began a programme of cleaning and re-dredging the canals in its urban stretches, with repairs to the floodgates to control adduction and drainage, with a view to remedying the problems of the flooded areas. At the same time, the state government, through the State Secretariat for Agriculture, Supply and Fisheries, carried out the redredging and cleaning of the Campos-Macaé and Coqueiros canals.

Since then, the Campos City Council's Environment and Civil Defence Department has implemented a periodic control and maintenance programme for the Coqueiros, Campos - Macaé, Rosário, Saco, Cacomanga, Jacaré, Vigário, São Bento, Quintinguta and Caxexa canals, among others, as well as activating the floodgates on the Flecha Canal.

Despite the interventions of the local government, due to the size and complexity of the system, the need for global planning was realised in order to implement rational use in the control of the water system and plan it to meet the potential demand from irrigated agriculture. With this concern, among others, in mind, the Baixo Paraíba do Sul Hydrographic Basin Committee (Baixo Paraíba do Sul Committee) was established on 3 March 2009 by State Decree No. 41,720. It operates in the Lower Paraíba do Sul Hydrographic Region, made up of the Muriaé, Pomba and Pirapitinga river basins and the lower course of the Paraíba do Sul river in the state of Rio de Janeiro.

The hydrographic region in which the Baixo Paraíba do Sul Committee operates (www.cbhbaixoparaiba.org.br) covers the municipalities of Quissamã, Natividade, São João da Barra, Cambuci, Itaperuna, São José de Ubá, Italva, Santo Antônio de Pádua, Cardoso Moreira, Aperibé, Miracema and Laje do Muriaé, as well as, partially, the municipalities of Trajano de Morais, Conceição de Macabu, Macaé, Carapebus, Varre-sai, São Francisco de Itabapoana, Campos dos Goytacazes, São Fidélis, Porciúncula and Santa Maria Madalena.

The structure of the Baixo Paraíba do Sul Committee consists of a plenary, the highest deliberative body, made up of 30 members with voting rights, ten of whom are representatives of Water Users, ten representatives of Civil Society and ten representatives of Public Authorities (federal, state and municipal). The collegiate board, made up of six members from the three segments that make up the Committee, is responsible for conducting its work.

The Committee also has three Technical Chambers, including the Technical Chamber for Water Resources and Hydraulic Structures, which studies problems relating to the resources, bodies of water and hydraulic structures within the Committee's area of activity or which have repercussions on it. Since then, this Committee, through the Technical Chamber for Water Resources and Hydraulic Structures, has discussed the recovery and maintenance of the region's water resources, seeking to reconcile the various interests of water resource users.

The Baixada Campista canal system set up by DNOS is fragile and complex, because despite its great length, the main canals have very low gradients (an average of less than 0.01%). This means that any level of silting or clogging along the canals prevents water from reaching rural areas, creating serious supply problems in the interior. Another consequence of clogged canals is that they also cause flooding in urban areas during the rainy season, causing great inconvenience to the population.

The problems arising from the system's malfunctioning are caused by the region's climatic conditions (water deficit), which have defined periods of rain and drought, as shown in Table 1.

Table 1 - Average climate data for the North of Rio de Janeiro.

months	temperature	precipitation	ET tank Class A	Relative humidity	wind speed	insolation	Eto (Penman)
	°C	mm	mm/day	%	km/h	hours	mm/day
January	27,1	134,0	5,1	76,1	2,4	6,9	5,6
February	27,7	82,0	5,5	73,7	1,9	6,8	6,0
March	26,1	83,0	4,6	75,9	1,7	6,2	4,9
April	24,0	83,0	3,8	77,7	1,5	6,4	4,0
May	22,3	54,0	3,1	77,4	1,4	6,2	3,4
June	21,4	34,0	2,6	77,2	1,5	6,4	2,9
July	21,4	42,0	2,9	74,3	1,8	6,9	3,0
August	22,1	30,0	3,5	76,5	2,2	5,3	3,7
September	22,9	57,0	3,7	77,7	2,5	4,5	4,1
October	24,3	99,0	4,1	77,9	2,5	5,1	4,6
November	25,3	153,0	4,7	78,2	2,4	4,8	5,2
December	25,1	172,0	4,7	75,6	2,2	4,8	5,1
average	27.1	85.3	4.0	76.5	2.0	5.9	4.4
total	-	1023,0	-	-	-	2143,0	1601,5

Source: Oliveira, 1996 - Master's thesis by Vicente de Paulo Santos de Oliveira - available at the Campos Centre Campus library - IFF

During the dry season, when the level of the Paraíba River is below 5.70 metres (DNOS), it becomes difficult to collect water for the canals and often it doesn't happen, making it difficult to regularise the water, which creates supply problems for rural properties, preventing their use for practices such as irrigation and animal watering, causing falls in agricultural and livestock production.

In the rainy season, there are problems of a different kind, because as the water level of the Paraíba rises, the floodgates need to be controlled in order to prevent flooding and damming in urban and rural areas.

Another major problem in the system is the maintenance of the canals, as the periodic control of the bottom levels requires equipment to be available for excavation and cleaning, as well as personnel to periodically control the aquatic vegetation and the slopes. In addition to these problems, the canals are also used for rubbish and sewage disposal, especially in urban areas, which exacerbates the problem.

From the table described above, it can be concluded that in the rainy season, due to the flat topography, the network of canals should play a predominant role in the drainage process, while in the dry season, they would act as inflow channels for agricultural use. The failure to regularise the system and interventions by individual interests mean that the system does not function properly, causing serious disruption in urban and rural areas, justifying the need for a management body to implement management and control projects of regional interest.

Three main morphological environments can be identified in the region: The Serra do Mar, geologically known as Cristalino; the Tabuleiros Terciários, often **called the "Sea of Hills" and the Baixada Campista, formed by the** alluvial **plain of** the Paraíba do Sul River.

The Baixada Campista is predominantly located on the right bank of the Paraíba do Sul River, a vast floodplain originating from an ancient bay which, after marine regression and continental rise, resulted in a large plain where it is common to find soils with clay layers whose distribution is chaotic. The explanation for this complex and heterogeneous distribution of soils on the Paraíba river plain can be explained by the geological history of this region, formed by marine and fluvial sediments from the Quaternary era.

The great spatial variability of the region's soils means that the **environment is also known as a "patchwork quilt", in reference to the** large number of classes that often occur in small areas, a marked characteristic of alluvial soils.

The low areas on the shores of the lagoons, located between the coastal ridges and the intertidal depressions, have a water table very close to the surface, often constituting bodies of water, where hydromorphs, often organic and semi-organic, predominate.

Among the bodies of water that occur in this area, Lagoa Feia stands out, where the Ururaí and Macabú rivers flow, occupying around 20,000 ha and connected to the sea by the Flecha Channel.

The sandbanks occupy areas between the Paraíba river plain and the Atlantic, being more prominent to the south of Lagoa Feia and in the region between Lagoa do Campelo to the north and Farol de São Tomé, where Quartz Sands predominate, which are of little interest from an agricultural point of view for irrigation purposes, although they are of great importance for maintaining the coastal ecosystem. In the municipality of São João da Barra, the lagoons of Grussaí, Iquipari, Salgada and Açu stand out, where the Port of Açu Logistics and Industrial Complex is currently being built.

Formed by lake and pond deposits, lake sediments are deposited in freshwater to slightly brackish environments, which are quite common in the area, where the characteristic sediment is a black-grey plastic clay rich in organic matter, and in swamps and marshes it can be formed by black organic material called peat.

9.5 Groundwater

Groundwater is a vital natural resource for the economic and safe supply of drinking water in urban and rural areas, which plays a fundamental role in the well-being of both human beings and many aquatic ecosystems (GW-MATE, 2001). It corresponds to the slowest part of the hydrological cycle and constitutes our main water reserve, occurring in volumes far greater than those available on the surface. In Brazil, most of our rivers only remain perennial because their base levels are maintained by aquifers in times of drought (MMA, 2001).

Most groundwater originates from excess rainfall that infiltrates the soil surface, so activities that take place on the surface can threaten its quality. Aquifer pollution occurs at points where the contaminant load generated underground by emissions and leachate produced by human activity is inadequately controlled and, in certain components, exceeds the natural attenuation capacity of soils and cover layers (GW-MATE, 2006).

Groundwater is found in saturated zones, where the pores, fractures or empty spaces in the solid matrix are completely filled with water. Therefore, as they are part of the same context, whatever happens to the soil will have repercussions on groundwater and may result in alterations to its quality. In this way, the migration of pollutants through the soil into surface and groundwater constitutes a threat to the quality of water resources used for public supply, industry, agriculture, commerce, leisure and services (CETESB, 2005).

Aquifers or natural groundwater reservoirs are rock formations or geological layers that store and transmit water with extraction capacity (CAPUTTI et al, 2001). According to the same authors, three types of aquifer can be differentiated based on the rock formation in which they are located:

- Granular or porous aquifers - those in which water is stored and flows in the spaces between the grains in sediments and sedimentary rocks with a granular structure. Examples: sandstones and alluvial fans.
- Fissure aquifers - those in which water is present in the fractures and cracks of crystalline rocks. Examples: granites, gneisses and diabases
- Karst or cavernous aquifers - those in which water is present in cavities produced by the dissolution of water. Examples: limestone and marble.

According to Albuquerque and Oliveira 1999, apud CAPUTTI et al., 2001, wells that are poorly located, built and operated without maintenance, the lack of government control over the opening of wells and the lack of basic hydrogeological studies can all be listed as risk factors for the use of groundwater.

From an environmental engineering perspective, the concept of water quality is much broader than simply characterising water by its molecular formula H_2O. This is because water, due to its solvent properties and its ability to transport particles, incorporates various impurities, which define water quality (VON SPERLING, 2006).

To characterise water. Various parameters are determined which represent its physical, chemical and biological characteristics. These parameters are indicators of water quality and constitute impurities when they reach values higher than those established for a given use (MOTA, 2003).

Water used in sanitary environments, changing rooms, kitchens, canteens, drinking fountains, safety equipment or in any domestic utility with direct human contact can be considered for human consumption (FIESP, 2004).

The quality of groundwater is determined, in principle, by the dissolution of minerals present in the rocks that make up the aquifers it percolates through. However, it can be influenced by other factors such as the composition of the recharge water, the contact time between water and the physical environment, the climate and even pollution caused by human activities (CAPUTTI et al, 2001). From a hydrogeological point of view, the quality of groundwater is just as important as its quantity. The availability of groundwater resources for certain types of use depends fundamentally on their physical-chemical, biological and radiological quality (FEITOSA; MANOEL FILHO, 1997).

In order to better identify the subdivisions of geomorphology in the Lower Paraíba do Sul region, the Rio de Janeiro Project (2001) highlighted the following formations:

- Tabuleiros de São Francisco de Itabapoana

This geomorphological unit occurs in a small portion of the study area, which corresponds to the left bank of the Paraíba do Sul River, in the districts of Travessão and Três Vendas. This relief is characterised by topographic amplitudes of up to 50m with very gentle gradients, rectilinear geometry, forming U-shaped valleys with flat and elongated tops, unconsolidated cover composed of colluvium and alluvium, low drainage density and parallel and dendritic patterns.

- Baixada Campista

Among the geomorphological units, the alluvial plains, represented by the Baixada Campista, correspond to most of the study area, including the urban area of Campos de Goytacazes. In this unit, the relief is characterised by the absence of gradients on the slopes, with geometry composed of sub-horizontal surfaces, the unconsolidated cover being formed by colluvial, alluvial and marine sandy-clay deposits and the occurrence of poorly drained terrain, which can be seen in the abundance of flooded stretches and the occurrence of artificial drainage channels.

- Bundles of Paraíba do Sul sandy strands

This unit only occurs in a small section of the eastern part of the studied area, making up the relief system of the coastal plains with a maximum elevation of 20 metres, slopes with a gentle to medium gradient, presenting undulating micro-relief due to the geometry of the ridges and the unconsolidated cover being made up of sandy sediments of marine and aeolian origin.

The geological characteristics of the state of Rio de Janeiro allow for the regional occurrence of two major aquifer systems, the fractured and the intergranular. The fractured aquifer system occupies most of Rio's territory, extending from the south to the north of the state through the open faults and fractures that occur in the rock masses. The intergranular aquifers, on the other hand, are made up of permeable sediments and alluvial and marine deposits that occur in the three main basins: Itaboraí, Resende and Campos. (Martins et al, 2007.

Capucci (2003) states that in 23 years of hydrogeological research and the construction of deep tube wells, the best results in terms of flow and water quality are related to the Campos Basin, in the southern portion of the Paraíba do Sul River. He also mentions the occurrence of four aquifer systems in Campos dos Goytacazes, the descriptions and characteristics of which can be found in Figures xxx.

The same author also suggests that the positive characteristics found in this region are related to the tectonic evolution of the Basin, where the great geological lineaments that occur in south-eastern Brazil have given rise to several NE/SW tilting faults, whose depth of occurrence has already been observed in up to 1963 metres of sedimentary package.

Caetano (2000) delineated four of the most representative aquifer systems in the region, three of which are sedimentary and one fractured: Fractured Aquifer, Tertiary Barreiras Formation, Tertiary Emboré Formation and Quaternary Deltaic, whose main characteristics are shown in Table 01.

Table 01: Aquifer formations in the Campos dos Goytacazes region

System	Composition	Specific Capacity (m³/h/m)	Flow (m³/h)
Fractured Aquifer	Pre Cambrian	0, 021 a 1,35	5,73
Barreiras Aquifer	Training Barriers	1,09 a 2,45	34,55 e 47,17
Emborê Aquifer	Emborê Formation	3,5	70,93
Quaternary Aquifer Deltaic	Residual soils, sands, gravels, clays and silts	90	139,67

Source: Adapted from Caetano (2000)

According to the Rio de Janeiro Project (CPRM, 2001), the main characteristics of existing aquifers are as follows:

- Fluviodeltaic Aquifer: Free aquifer, located to the south-west of the city of campos, on the right bank of the Paraíba do Sul River, in an area of approximately 304 km2. It is made up of outcropping sediments of Quaternary age, composed of fine- to medium-grained sands and sandstones, with a silty matrix and clay bands, with thicknesses varying between 60 and 90 metres, overlying the crystalline basement and older sediments. The water is of good quality, occasionally ferruginous.

- Emborê Aquifer: This is a confined aquifer, made up of non-outcropping sediments of possible Tertiary age, made up of varied conchiferous sandstones, with feldspar; impure claystones and organic clays, with thicknesses of up to 200m. It is located near Farol de São Tomé and covers an area of approximately 350 km². The water is of good quality and usually has a low total dissolved solids content.

- São Tomé II Aquifer: This is a confined aquifer, completely covered by quaternary sediments, with a thickness of approximately 230m, and in the vicinity of Farol de São Tomé it can reach up to 2,000m. It is located throughout the eastern part of the Campos Basin, in an area of approximately 910 km². It is composed of reddish, lateritic sandstones with calcium clays, over consolidated clayey sandstones. This aquifer is usually found between 80 and 160 metres deep. There are ferruginous waters with total dissolved solids between 200 and 600 mg/l.

- São Tomé I Aquifer: This aquifer covers an area of approximately 380 km², forming an elongated strip running NE-SW in the central part of the Campos Basin. It is made up of non-outcropping sediments of Tertiary age, completely covered by Quaternary sediments. This system is made up of reddish, lateritic sandstones with calcitic clays, over consolidated clayey sandstones. It is a confined aquifer up to 160 metres thick, overlying the crystalline basement. In terms of quality, ferruginous water can occur and STD ranges from 500 to 1,800 mg/l.

- Barreiras Aquifer: This aquifer covers an area of approximately 1,630 km² and is located on the western edge of the Campos Basin. It consists of outcropping Tertiary sediments made up of lateritic clays and sands with iron oxide. It is free-flowing and usually has ferruginous waters.

CHAPTER 10

AQUACULTURE AND SUSTAINABILITY

Luiz Paulo de Oliveira Tostes

Felipe Cabral da Costa de Barros

Milton Erthal Junior

Ana Paula Ribeiro Costa Erthal

Osmane Pessanha Ribeiro

Henrique Monteiro da Hora

Alcimar Ribeiro Chagas

10.1 SUSTAINABLE DEVELOPMENT AND AQUACULTURE

One of the advantages of aquaculture in the modern world is related to the paradigm of environmental sustainability. This activity has a direct impact on rivers and river basins, as well as contributing to the development of a more sustainable rural environment. Some studies show that aquaculture, with its various functions, can play a fundamental role in the hinterland because, as well as producing fish to restore natural stocks in overexploited rivers, aquaculture has generated work and income for riverside populations, increasing the sustainability of rural systems by integrating with other productive activities. FAO documents also show that in recent decades aquaculture has been integrated into the economies of many third world countries, increasingly contributing to the generation of foreign currency and food security in rural areas.

One of the FAO's most important instruments for the fisheries and aquaculture sector is the "Code of Conduct for Responsible Fisheries" (FAO, 1995). The need for effective mechanisms to regulate international fisheries arose at the end of the 1980s, when it became clear that fisheries resources would not support exploitation and development on the same scale and speed as in previous years. This Code establishes international principles and standards for the application of responsible practices, aimed at ensuring the conservation, management and effective development of living aquatic resources, with respect for the ecosystem and the environment.

The Code recognises the nutritional, economic, social, cultural and environmental importance of fishing. According to the Code, the term "fisheries" also includes aquaculture and the interests of all those who relate to the fisheries sector, taking into account the biological characteristics of the resources, the environment and the interests of consumers and other users. The Code is a point of reference for national and international initiatives that seek a sustainable exploitation of living aquatic resources, in harmony with the environment. Although its application is not mandatory, some of its parts are based on international law standards. It also contains provisions that can be granted or have binding effect. It contains mandatory legal instruments relating to the fisheries sector, as well as links to other international instruments, in particular the "Rio Declaration on Environment and Development" and "Agenda 21", adopted by the 1992 United Nations Conference on Environment and Development (UNCED). The Code is applied worldwide, to FAO members and non-members, to governmental and non-governmental entities, to regional and global organisations, in short, to all those involved in fishing and the aquatic environment.

Among the main issues raised by the FAO for the aquaculture sector is the need to create an environment suitable for sustainable aquaculture. In the report *"The state of world fisheries and aquaculture 1998"* (FAO, 1999), this issue is addressed in the following terms: "It is considered that few countries have an appropriate political and legal framework for aquaculture and that these aspects have been neglected because its development has been seen as essentially technical and, consequently, the assistance given has been more focussed on the technical aspects of production. Neglecting to incorporate political, economic, social, environmental and legal aspects often has negative consequences for the sector.

From the FAO's perspective, the diversity of the sector is a guarantee of the many benefits that aquaculture can bring. This diversity involves food, inputs, direct and indirect jobs and social opportunities. Among these, according to Muir and Nugent (1995), food security is pointed out as one of the most important, since it provides food directly to the producer or to the communities in the immediate vicinity, as is the case in many areas of Asia. This contribution can also be indirect, such as a more regular and reliable economic activity, especially when compared to traditional fish catching. It is also an option for diversification, creating new opportunities for rural producers. In small-scale aquaculture, the involvement of families and communities is common. For specific development policies, this offers various possibilities, such as: the involvement of disadvantaged groups; activities aimed at benefiting women; or helping settled groups in areas where water bodies are underutilised.

The concepts of sustainability are relatively new and there are few concrete examples of sustainable aquaculture. The majority of aquaculture development models in Brazil are not the product of proper planning, largely due to the existence of models that only take into account the economic aspect, devoid of environmental and social concerns (VINATEAE, 1998). However, sustainable aquaculture should be understood as the viable production of aquatic organisms over time (PILLAY, 1996). This model is based on three basic concepts: economic efficiency, ecological prudence and social equity (VINATEAE, 1998).

The sustainable use of natural resources, especially fisheries resources, is a challenge that needs to be tackled from a technical, political, economic and social point of view. The future of the population of developing countries will depend on the balance between the exploitation and conservation of natural resources. Food production and the livelihoods of food producers and workers associated with the industry will have to guarantee a **"sustainable"** system **rather than an overexploited environment with changes to its** normal operating pattern. According to Pullim (1989), the environment should be defined broadly, in which human ecology is related to biotic and abiotic natural resources.

Compared to agriculture, aquaculture is a recent practice and, although the merits of aquaculture in terms of food production and as a hub for integration with agriculture are recognised, the impacts on the natural systems that receive its effluents have also been acknowledged. Furthermore, it must be considered that the aquatic environment supports a series of physical, chemical and biological processes which, together with the metabolic activity of the farmed fish themselves, are extremely important for the sustainability of the environment (CONTE, 2002).

Modern aquaculture is based on three pillars: profitable production, environmental preservation and social development. The three components are essential and inseparable if we are to have a perennial activity. It must therefore be understood that environmental preservation is part of the production process (VALENTI et al., 2000).

Ecological concerns have become an important variable for public administrations to consider. Pressure from international organisations, the government and society to improve environmental quality has led municipal authorities to seek answers to these demands. For a long time, responsibility for environmental policies was centralised in the hands of state and federal bodies. Since CONAMA resolution 237/97, the assessment of local environmental impacts caused by projects has also become the responsibility of the municipality. However, the municipal government's main responsibility is to coordinate actions and develop, together with its community, a coherent environmental mindset, with a view to implementing rules to control environmental deterioration and seek the necessary rehabilitation of the most affected areas. To this end, it must fully assume its mission of guiding the sustainable development of its community, based on criteria of social equity, economic development and environmental protection. It should also be emphasised that it is the duty of the supervisory bodies to ensure the integrity of the environment by more rigidly assessing implementation projects that involve manipulation of any part of the river basin, such as the breeding of aquatic organisms (ARANA, 1999). Due to the diversity of the production systems used and the different perceptions of the concept of **"sustainability", it is necessary to use a more balanced and** scientifically based **strategy** to discuss issues relating to the development of this activity and its consequences for the environment. To this end, it is essential that there is a mutual commitment between the various actors that make up the aquaculture production chain in order to enable its sustainable development.

However, what is most needed are concrete experiences of sustainable development that allow us to establish a benchmark for evaluating aquaculture. Actions aimed at such development should favour forms of intervention that, in addition to economic development, guarantee the maintenance of natural resources for the enjoyment of future generations.

10.2 ENVIRONMENTAL IMPACTS

Fish production in Brazil has grown even more than that seen on a global scale. According to Segalin (2007), fish farming grew by a significant 275% between 1996 and 2005, but the activity also faces environmental problems, especially in terms of effluent emissions. As well as requiring the use of quality water to cultivate aquatic organisms, this activity generates effluents rich in nutrients and parasites that cause environmental degradation (ZANIBONI, 2005). These problems are also highlighted by Bastian (1991) who mentions that fish farming depends on the use of water free from pollutants, yet fish farming is an activity that causes potential environmental degradation.

The impact of fish farm effluents on receiving water bodies depends on the difference in concentration of water quality parameters between pond water and stream water, as well as the proportion of water originating from the pond and the stream (BOYD, 1990). In this way, the polluting potential of a given cultivation system can vary depending on where it is installed. The characteristics of the effluent generated by fish farming are very similar to those of domestic effluent, i.e. it has a high biochemical oxygen demand and a high concentration of suspended solids, nitrogen and phosphate compounds. This similarity allows for an analogy of the impacts caused by crops, contributing to the eutrophication of rivers, streams and lakes (ZANIBONI, 2005).

Aquaculture effluents are rich in nitrogen, phosphorus and organic matter, which contribute to the eutrophication of receiving water bodies, as well as reducing or altering biodiversity (HENRY-SILVA and CAMARGO, 2008).

The excessive introduction of organic matter into the water, whether in the form of feed, manure or agro-industrial by-products, allows an exaggerated development of phytoplankton, which leads to an ecological imbalance, depression of oxygen concentrations and asphyxiation of aerobic beings (BRITSKI **et al.**, 1972). In addition to the high proliferation of algae, the excess of nitrogen compounds, in the form of ammonia, nitrite and nitrate, directly affects fish. According to Sipaúba-Tavares (1995), these compounds raise the pH of the blood, affect the osmotic exchange of fish and reduce the internal concentration of ions, increasing oxygen consumption in the tissues, damaging the gills and reducing the ability of the blood to transport oxygen. This causes histological changes, especially in the kidneys and spleen, and increases susceptibility to disease (ARANA, 1997).

Another problem is the introduction of exotic species. Breeding fish from other regions of the country or even from other countries can cause local and regional environmental impacts when the species in question reaches the region's water sources. These fish can carry diseases or pests which, when transmitted to native fauna that has not evolved with the aetiological agent in question, can cause imbalances in native populations or even eliminate them. In addition, exotic species can be accidentally introduced into the region's rivers, streams and lakes and compete with the native fauna for food and shelter.

Many fish farmers, by building their ponds irresponsibly, cause environmental impacts by breaking their dykes. Ruptured dykes cause fish kills in rivers and streams on the way down due to physical (temperature, suspended solids) and chemical (pH, nutrients) factors in the water. Another important factor is the introduction of exotic species by uninformed farmers. During these accidents, these animals can reach regional rivers and lakes, jeopardising the local fauna.

Recently, we've seen reports of fishermen catching these animals in the region's rivers and lagoons. The African catfish and tilapia are constant examples of exotic fish that can be caught in the Norte Fluminense region. The African catfish, being a predatory and very voracious fish, can eliminate a large number of native fish in order to develop. In a recent story (July 2012), the local press in Campos dos Goytacazes reported the capture of a pirarucu, a carnivorous fish from the Amazon region, in Lagoa de Cima, which was over 2 metres long. Cultivation of this species is prohibited by the environmental agencies in the region (INEA and IBAMA), as it can cause serious imbalances in native fish populations.

REFERENCES

ARANA, L. V.; Chemical principles of water quality in aquaculture: a review for fish and shrimp. Florianópolis: UFSC, 1997. 166 p.

ASSUNÇÃO, A. W. A.; Treatment of fish farm effluents using a wetland system populated with aquatic macrophyte species of three different ecological types. 2011. 71 f. Dissertation (Master's Degree) - Postgraduate Course in Continental Water Aquaculture, Unesp Aquaculture Centre, Universidade Estadual Paulista, Jaboticabal, 2011

AZZOLINI, J. C.; ZARDO, F; SEGALIN, C. A.; Monitoring of effluents from fish pond discharge at different levels of discharge. Unoesc & Ciência- Acet, Joaçaba, v. 1, n. 1, p.59-68, 2010.

NATIONAL DEVELOPMENT BANK. Panorama of aquaculture in Brazil: challenges and opportunities: Agroindustry. BNDES Sectoral 35, p. 421 - 463. Available at: <www.bndes.gov.br/ >. Accessed on: 01 April 2012.

BARROS, F. G. N.; AMIN, M. M.; Water: scarcity, valuation and prospects for brazil. Available at: <http://www.sober.org.br/palestra/2/162.pdf>. Accessed on: 26 March 2012

CAMPANILI, M.; In Brazil, there is a deficit in the midst of abundance. São Paulo: Agência Estado, Caderno Ciência, 2003. Available at http://www.estadao.com.br/ext/ciencia/agua/aguanobrasil. Accessed on 23 Feb 2005.

CONNOLLY, K; TREBIC, T.; Optimisation of a backyard aquaponic food production system. Faculty of Agricultural And Environmental Sciences Macdonald Campus, Mcgill University: Bree 495 Design 3 Bioresource Engineering Presented To Dr Vijaya Raghavan, 2010. 74 p.

DIVER, S.; Aquaponics - integration of hydroponics with aquaculture. ATTRA - National Sustainable Agriculture Information Service, Fayetteville, AR, USA. 28 pp. 2010

DUARTE, E.; Cultivation of nile tilapia post-larvae using different proportions of shell/gravel substrate in the biofilter. 2010. 47 f. Dissertation (Master's Degree) - Animal Science Course, Universidade Federal Dos Vales do Jequitinhonha e Mucuri, Diamantina, 2010.

FONSECA, J.P; FONSECA, C. F. A; REIS, R. F; FREITAS, J. Corporate environmental management applied to the case study of water treatment in new york. 2009. 85 f. Monograph (Bachelor's Degree) - Business Administration Course, Centro Universitário Metropolitano de São Paulo, Guarulhos, 2009. Available at: <http://www.folgueral.com.br/anexo/tcc/tcc_agua.pdf>. Accessed on: 25 Feb. 2012

GIA. Sector study to consolidate sustainable aquaculture in Brazil: 2007. Available at at: <ftp://ftp.fao.org/fi/document/aquaculture/sect_study_brazil.pdf>. Accessed on: 23 Apr. 2012.

HENRY-SILVA, G. G.; CAMARGO, A. F. M.; Impact of aquaculture activities and effluent treatment systems on aquatic macrophytes - case report. Revista Científica de Pesca, Aquicultura e Limnologia, São Paulo, v 34, n.1 , p.163-173, 22 Aug. 2007.

LEMOS, H. M.; The 21st Century and the Water Crisis. São Paulo: Agência Estado, Caderno Ciência, 2003. Available at http://ambientebrasil.com.br. Accessed on 05 Feb 2012

MACEDO, C. F.; SIPAÚBA-TAVARES, L. H.; Eutrophication and water quality in fish farming: consequences and recommendations. Bol. Inst. Pesca, São Paulo, v. 2, n. 36,p.149-163, 04 Nov. 2010. Available at: <ftp://ftp.sp.gov.br/ftppesca/36_2_149-163rev.pdf>. Accessed on: 07 Feb. 2012.

MASUDA, C. T.; Trends and prospects for fish production in Brazil. 2009. 55 f. Monograph (Bachelor's Degree) - Veterinary Medicine Course, Faculdade Metropolitanas Unidas, São Paulo, 2009

MPA. Fisheries and aquaculture statistical bulletin: Brazil 2010. Available at: <www.mpa.gov.br/ >. Accessed on: 29 January 2012.

NELSON, R. L.; PADE, J. S.. Aquaponic food production. Montillo, Wi. USA. 2008. 218 p

NOGUEIRA FILHO, Hercules et al. Aquaponics: interaction between hydroponic lettuce and super-intensive Tilapia farming.Available at: <www.abhorticultura.com.br/biblioteca/arquivos/.../cpfg2025c.pdf >. Accessed on: 12 Feb. 2012.

OLIVEIRA, R. C.; O panorama da aquicultura no brasil: a prática com foco na sustentabilidade. Revista Intertox de Toxicologia, Risco Ambiental e Sociedade, São Paulo, v. 2, n. 1, p.71-89, 01 feb. 2009.

RAKOCY, J. E., MASSER, M. P., & LOSORDO, T. M. . Recirculating Aquaculture Tank Production Systems: Aquaponics-Integrating Fish and Plant Culture. Southern Regional Aquaculture Centre, Publication No. 454 , 1-16. 2006

RAKOCY, J.E.; HARGREAVES, J.A. Integration of vegetable hydroponics with fish culture: a review. Techniques for Modern Aquaculture, SRAC Publication No. 453 p.112-136,1993

REBOUÇAS, B. B., CUNHA. A., BRAGA B. Águas doces no Brasil: capital ecológico, uso e conservação. São Paulo. 1999.

SEBRAE. Santa Catarina in numbers: fishing and aquaculture - 2010. Available at: <www.sebrae.com.br>. Accessed on: 23 March 2012.

SHIMODA, E.; ANDRADE, D.R.; Fish farming: water quality monitoring and control practices. Technical Bulletin - UENF. v 3. n 3 Campos dos Goytacazes. 1998.

SOUZA, J. A. P. L. L.; Study of the social, economic and environmental impacts of fish farming in net-tanks in the region of paulo afonso- ba. 2006. 131 f. Dissertation (Master's) - Environmental Sciences Course, Federal University of Tocantins, Palmas, 2006.

UNIAGUA. Water on the planet. Available at: < http://www.uniagua.org.br/website/default.asp?tp=3&pag=aguaplaneta.htm>. Accessed on: 06/04/2012

VALENTI, W. C. 2002. Sustainable aquaculture. In: Zootechnics Congress, 12th, Vila Real, Portugal, 2002, Vila Real: Portuguese Association of Zootechnical Engineers. Proceedings...p.111-118.

WERLANG, S. R. C. The importance of sustainability. Electronic Review 21/01/2008.   Available at:

<http://www.fazenda.gov.br/resenhaeletronica/MostraMateria.asp?page=&cod=4 34831>.
Accessed on: 14 May 2012.

ZANIBONI FILHO, E. Treatment of fish farm effluents. Annals of ZOOTEC 2005 - 24 to 27 May 2005 - Campo Grande-MS. Available at: <www.lapad.ufsc.br/PDF/pdf0014.pdf>. Accessed on: 18 April 2012.

CHAPTER 11

ENVIRONMENTAL LICENSING OF CONTINENTAL AQUACULTURE

Osmane Pessanha Ribeiro

Forestry Engineer

Head of Environmental Licensing INEA - Campos dos Goytacazes R.J. Professor of Secondary, Technical and Technological Education IFF Campus Guarus - Campos dos Goytacazes R.J.

11.1. INTRODUCTION

Aquaculture, especially fish farming, represents a profitable economic activity for rural producers, requiring technical knowledge of the biology and management of the species they wish to cultivate, as well as knowledge of the activity's influence on the ecosystem in which the property is located, so that negative environmental impacts are minimised or even eliminated from the production process.

Considering this context, it is essential to adopt ecologically viable techniques, from the choice of species suitable for confinement in excavated tanks, through the rational use of water resources, to the disposal of wastewater in the environment.

In this chapter, we will focus on the legal aspects surrounding the environmental licensing of aquaculture activities, focussing on the compatibility of economic exploitation with sustainability and environmental protection.

It should be noted that the administrative procedures involved in obtaining the relevant environmental licence do not create barriers to production. The projects and studies requested by the environmental control body in a way lead the aquaculture farmer to develop his activity rationally, achieving better yields and, at the same time, greater respect for the environment in which he lives.

11.2. ENVIRONMENTAL LICENSING OF CONTINENTAL AQUACULTURE IN THE STATE OF RIO DE JANEIRO

11.2.1. Considerations. Initial

The National Environmental Policy (Federal Law No. 6.938 of 1981) defines pollution as the degradation of environmental quality resulting from activities that directly or indirectly:

a) Harm the health, safety and well-being of the population;

b) Create adverse conditions for social and economic activities;

c) Affect the biota unfavourably;

d) Affect the aesthetic or sanitary conditions of the environment;

e) Release materials or energy in breach of established environmental standards.

Aquaculture is a polluting activity in that it uses water in the production system, generating solid waste and contaminated liquid effluents that need to be treated before being disposed of in the environment.

SISNAMA (the National Environmental System), provided for by the National Environmental Policy, created the hierarchy of environmental licensing in Brazil. At the federal level, the Brazilian Institute for the Environment and Renewable Natural Resources (IBAMA) is now the licensing body for activities that actually and/or potentially pollute, while the sectional (state) and local (municipal) bodies deal with the issue in their respective administrative spheres.

With regard to the use of water resources, the abstraction of surface or underground water for the activity in question requires the acquisition of a document granting the right of use for the appropriate purposes. Federal Law No. 9.433 of 1997, which deals with the National Water Resources Policy, and Federal Law No. 9.984 of 2000 talk about specific public policies and basic procedures for granting licences, respectively, within the context that water represents a common good for all, finite in nature and endowed with economic value.

The National Environment Council (CONAMA), the federal advisory and deliberative body that makes up SISNAMA, promulgated Resolution No. 413 on 26 June 2009, which provides for the environmental licensing of aquaculture in Brazil and serves as a technical instruction for aquaculture producers to regularise their activity.

In the state of Rio de Janeiro, given its constitutional hierarchy and its consequent openness to the possibilities of regulatory restrictions associated with speeding up requests for licences for aquaculture activities, the state environmental licensing and control body (INEA - Instituto Estadual do Ambiente) has the Environmental Licensing System (SLAM), enacted in State Decree No. 42.159 of 2009, replaced by State Decree No. 44.820 of 2014 (updated by State Decree No. 45.482/2015).

The environmental licensing of inland aquaculture activities basically involves two stages according to the following licence modalities:

□ Preliminary and Installation Licence (LPI): administrative act by which the environmental agency, in the preliminary planning phase of the undertaking or activity, approves its location and design,   certifying the environmental viability,   whileat the same time as authorising   a installation of the undertaking or activity in accordance with the specifications contained in the approved plans, programmes and projects, including the environmental control measures and other conditions, of which they constitute the determining reason.

The IPL may authorise pre-operation, for a period specified in the licence, in order to obtain the data and performance elements needed to support the granting of the Operating Licence.

Operating Licence (LO): an administrative act by which the environmental agency authorises the operation of an activity or enterprise, after verifying effective compliance with the provisions of previous licences, with the environmental control measures and other conditions determined for the operation.

Alternatively, SLAM also provides for the following case of a single phase for the activity in question, depending on the degree of impact and local and project particularities:

Simplified Environmental Licence (LAS): an administrative act by which the environmental agency, in a single phase, certifies the environmental viability, approves the location and authorises the implementation and/or operation of undertakings or activities that fall into the low impact classes, establishing the conditions and environmental control measures that must be observed.

The initial step-by-step instructions for applying for a licence, as well as the definition of the type of licence, will depend on the legal and technical framework to be set up on INEA's Environmental Licensing Portal.

The State Secretariat for the Environment (SEA), through the Rio de Janeiro State Environment Council (CONEMA), at its meeting on 2 August 2013, using the powers conferred on it by State Decree 40.744 of 25 April 2007, approved CONEMA Resolution 49, published in the D.O.U. of 14 August 13, which provides for the revision of the Operational Standard of State Institute for the Environment No. 04 - Licensing of Continental Aquaculture (NOP-INEA- 04).

The purpose of NOP-INEA-04 is to establish criteria and procedures to be adopted in the state of Rio de Janeiro for the environmental licensing of

enterprises geared towards continental aquaculture activities, in extensive, semi-intensive or intensive systems, in different cultivation structures. However, shrimp farms are excluded from this rule.

NOP-INEA-04 is applied to all applications for environmental licences for continental aquaculture projects, whether new or already underway at the State Environment Institute (INEA), including processes that are being renewed and for which the required licence has not yet been issued.

Concepts defined by NOP-INEA-04

Continental aquaculture: the cultivation or breeding of organisms whose life cycle, under natural conditions, takes place wholly or partially in a freshwater environment.

Artificial environments: reservoirs and weirs.

Exotic species: species originating from another country.

Allochthonous species: species that does not occur or has not occurred naturally in the Geographical Reference Unit (GRU) in question Native species: species of origin and natural occurrence in waters of the Geographical Reference Unit in question.

Geographical Reference Unit (GRU): the area covered by a hydrographical region, or in the case of marine and estuarine waters, strips of coastal waters between two points on the Brazilian coast.

Young forms: fry, tadpoles, imagoes, larvae, seaweed seedlings intended for breeding, nauplii, eggs, post-larvae and seeds of bivalve molluscs.

Extensive cultivation system: the production of cultivated species depends above all on the natural food available, with the possibility of receiving artificial food in addition, and is characterised by a medium or low density of specimens, which varies according to the species used.

Intensive cultivation system: production in which the cultivated species depend entirely on the supply of artificial food, one of the characteristics of which is the high density of specimens, varying according to the species used.

Semi-intensive cultivation system: production in which the cultivated species depend especially on the supply of artificial food, and may seek to complement the natural food available, and which is characterised by a medium or low density of specimens, varying according to the species used.

Barrage Structure: built in the channel of a watercourse, perpendicular to its flow, designed to contain/brake the natural flow of water.

11.2.2. *Criteria for Environmental Licensing*

The definition of size and polluting potential for aquaculture enterprises will be in accordance with the criteria defined in INEA Resolution No. 52/2012, which establishes the new codes for the classification of enterprises and activities that pollute or use environmental resources, as well as those capable of causing environmental degradation, subject to environmental licensing, and Resolution No. 53/2012, which establishes the new criteria for determining the size and polluting potential of enterprises and activities that pollute or use environmental resources, as well as those capable of causing environmental degradation, subject to environmental licensing.

The most restrictive criterion in environmental terms will be taken into account when classifying aquaculture enterprises that raise several species.

The cost of analysing the environmental licence application will be calculated based on the table in Annex 02 of NOP-INEA-02, approved by CONEMA Resolution 31/2011.

In the environmental licensing of continental aquaculture projects in permanent preservation areas, the provisions of CONEMA Resolution 48/2013 and the other legal provisions in force must be observed. If the project is located in areas with native vegetation from the Atlantic Forest Biome, it must comply with the provisions of Law No. 11.428/2006 and Decree No. 6.660/2008, which regulates it. In the case of areas located in sustainable use conservation units or in the buffer zone of integral protection units, their respective management plans, specific legislation and, in particular, CONAMA Resolution 428/2010 must be observed.

For the closure of aquaculture activities, a Deactivation and Recovery Plan must be submitted to the environmental agency, with the appropriate implementation schedule.

Documents for applying for environmental licences for continental aquaculture enterprises in accordance with NOP-INEA-04

General documents

1. Application form completed and signed by the legal representative (Form n°01 available on the INEA website).

2. Declaration of delivery of documents in printed and digital form (Form n°02, available on the INEA website).

3. Copies of the ID and CPF documents of the legal representative signing the application. If the applicant is an individual, they must also provide proof of residence.

4. If there is a proxy, submit a copy of the public or private power of attorney with a notarised signature, and copies of the identity and CPF documents. Copies of the identity documents and CPF of the INEA Contact, indicated by the legal representative.

5. Copy of the minutes of incorporation and election of the last board of directors and bylaws, in the case of a S/A, or updated articles of association in the case of a limited liability company. If the applicant is a public body, the Act of Appointment of the legal representative signing the application must be submitted.

6. Copy of registration in the National Register of Legal Entities (CNPJ).

7. Copy of the Certificate from the City Council stating that the site and the type of enterprise or activity are in compliance with the legislation applicable to land use and occupation.

8. Copy of the property title and updated certificate from the General Property Registry (RGI); or copy of the leasing certificate, if applicable; or copy of the Assignment of Use, in the case of Federal/State property; or Declaration of Ownership, if applicable. If the applicant does not own the property, they must also submit a Lease, Commodity Contract or other agreement (Optional in the case of a Prior Licence - LP).

9. If the property is rural, the Rural Environmental Registry (CAR) must be presented.

10. Copy of the CPF number and registration with the professional council of the person(s) responsible for the project, construction or operation, accompanied by an up-to-date Technical Responsibility Certificate (ART).

11. Location plan, using copies of IBGE plans, Google Earth maps, sketches or others, indicating:
    - UTM or geographic coordinates;
    - Location of the land in relation to the main road and at least two others, indicating the name of the accesses; if it is located on the side of a road or motorway, indicate the kilometre and the side on which it is located;
    - Bodies of water (rivers, lakes, etc.) closest to the development, with their respective names, if any;
    - Uses of neighbouring properties and areas within a radius of at least 100 metres.

Specific Documents According to NOP-INEA-04

1. Use of raw water must be conditional on obtaining a Water Resources Use Right Grant or an Environmental Certificate of Insignificant Use, the latter in the case of groundwater extraction of less than 5,000 litres per day or surface water abstraction of less than 34,560 litres per day and a flow rate of less than 0.4 litres per second.

2. If the water to be used is the domain of the Federal Government, the licence must be obtained from the National Water Agency (ANA).

3. If effluents from the activity are discharged into water sources, a water use licence must be obtained to discharge the effluents, which must meet the effluent discharge standards listed below, established in CONAMA Resolution 357/2005, supplemented and amended by

CONAMA Resolution 410/2009 and CONAMA Resolution 430/2011, and in Technical Standard NT-202.R-10, which establishes criteria and standards for the discharge of liquid effluents, a revision approved by CECA Resolution No. 1007 of 04 December 1986 and published in the Official Gazette of 12 December 1986. The main parameters to be checked:

- pH between 5.0 and 9.0.
- Sedimentable materials up to 1.0 ml/L in a one-hour Imhoff Cone test.

  M Floating materials virtually absent.
- Ammoniacal nitrogen up to 5.0 mg/L.

4. If the effluent is discharged into lakes, lagoons, lagoons or reservoirs, **or into a watercourse that contributes to one of these bodies of water,** the discharge standards for the parameters total nitrogen (10.0 mg/L) and total phosphorus (1.0 mg/L) **must be** met, in accordance with Technical Standard NT-202.R-10, which establishes criteria and standards for discharging liquid effluents, a revision approved by CECA Decision No. 1007 of 04.12.1986 and published in the Official Gazette of 12.12.1986.

5. Effluents discharged directly into lakes, ponds, lagoons or reservoirs must be free of sedimentable materials in a one-hour Imhoff Cone test.

6. If the effluent is destined for the public sewage system, the local sewage treatment concessionaire must give its consent.

7. The use of chemical products in cultivation must be informed and justified, and will be duly analysed and authorised by the environmental agency as part of the licensing process, by presenting control measures.

8. Registration of Aquaculture Enterprises, duly completed and signed by the legal representative (Annex - 01 of NOP-INEA-04).

9. Proof of Registration in the Federal Technical Register (for licence applications environmental for new activities or activities that already exist and are in the process of obtaining their first licence) and Certificate of Good Standing in the Federal Technical Register of Polluting Activities (IBAMA) (for activities that already have an environmental licence).

10. Consent from the managing body of the conservation unit, where applicable.

11. Technical draft of the project, accompanied by an annotation or registration of technical responsibility (presented at the initial stage of the project, in the case of a request for a preliminary licence).

12. Descriptive memorial, along the lines of the Environmental Report in CONAMA Resolution 413/2009, as amended by CONAMA Resolution 459/2013.

13. Situation plan of the development area, on an appropriate scale, indicating interventions in Permanent Preservation Areas.

14. Technical project for the development, accompanied by a technical responsibility annotation or registration.
15. Environmental monitoring programme in accordance with CONAMA Resolution No. 413/2009 as amended by CONAMA Resolution No. 459/2013 (submitted at the application stage for an operating licence or simplified environmental licence).
16. Registration in the National Water Resources Register (presented at the application stage for an operating licence or simplified environmental licence).

Criteria to be considered when drawing up the technical project according to NOP-INEA-04

- If the project uses raw water, the following criteria must be met:

  - Provide a screen or filter to prevent the introduction of unwanted aquatic organisms.
  - The catchment by direct derivation must have a sluice and thief to control the flow and avoid problems arising from flooding.
  - In the case of catchment by bypass channels, these must be sized and **protected so that there is no infiltration or overflow of water, and** the vegetation on their banks must be maintained in order to prevent erosion.
  - The nurseries must have suitable drainage structures, such as **a monk, bit pipe and elbow.**
  - The facilities must have mechanisms to prevent and control the escape of farmed animals throughout the production process.

- In the event of infectious diseases, the breeding and quarantine facilities must be isolated and the water disinfected before being released into the environment.
- If the water from the cultivation tanks is to be reused, it must go through physical and biological filtration processes before being reused.
- Pesticides may not be used in the vicinity of water supplies, tanks or canals.
- The use of chemical products in cultivation must be informed and justified, and will be duly analysed and authorised by the competent environmental body as part of the licensing process, by presenting control measures.
— Crops grown in bodies of water owned by the Federal Government must comply with Interministerial Normative Instruction 06/04.
— Aquaculture projects that require the construction of new ponds by damming the river channel will not be licensed.
— Aquaculture projects in lakes, ponds and lagoons will not be authorised.
— Continental aquaculture ventures may not be set up on land located less than 15 metres from roads, railways and pipelines, or in their right-of-way.

— The selection of species for breeding must comply with the restrictions laid down in IBAMA Ordinance No. 145/98, which establishes rules for the introduction, reintroduction and transfer of fish, crustaceans, molluscs and aquatic macrophytes for aquaculture purposes.

— The feed to be used to feed the species must be stored on a structure at least 40 cm high, away from the wall, so that it can be inspected all around to check for the presence of vectors (harmful insects and rodents) and other animals that could contaminate the feed.

— The release and introduction into the natural environment of exotic and alien species is prohibited, in accordance with Article 9 of IBAMA Ordinance No. 145/98.

— The release and introduction into the natural environment of native animals bred in captivity will depend on authorisation from the state environmental agency, based on a study of population dynamics.

— If animals die, they must be incinerated or buried in pits containing lime, and cannot be disposed of in the environment or sent to urban waste disposal sites.

Criteria for environmental licensing of fish farms in excavated ponds, lined tanks and raceways in accordance with NOP-INEA-04.

- Fish farming projects in excavated ponds must provide for a distance of at least 50 cm between the bottom of the pond and the water table, the depth of which must be specified when applying for the environmental licence, taking into account the type of substrate and the time of year.

- In fish farms that have already been set up, where the water table is rising and there is no water renewal, the biomass in the ponds should not exceed 200 g/m^3.

- Higher concentrations of biomass may be approved, provided that the adoption of mechanisms and procedures capable of protecting the water table is proven.

- Where the impact has been classified as medium or high, the documentation for applying for the environmental licence must include the granulometric profile of the soil, according to the depth of the planned nursery. In cases where the clay content is less than 35%, the land must be waterproofed.

- The construction of slopes and nurseries must meet the following criteria:

  • The soil used for construction and for lining the bottom of the ponds must not contain organic matter, stones or other materials that allow infiltration and/or are too porous.

  • The slope must maintain a ratio of 2:1 between its height and the horizontal projection on the inside and 1:1 on the outside. See Figure 1.0.

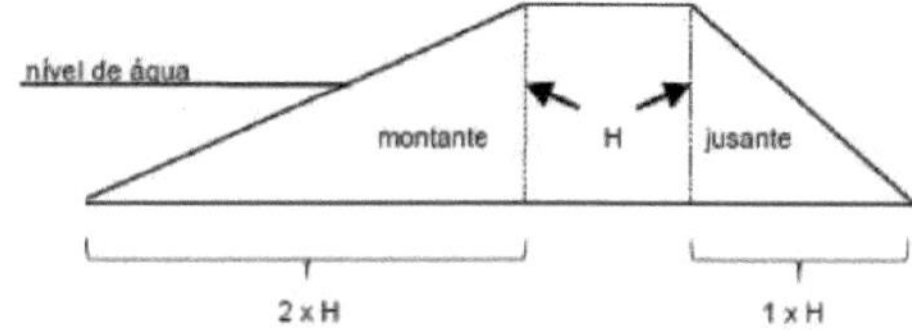

Figure 1.0 Ratio of 2:1 between its height and the horizontal projection, on the internal face, and 1:1. Source: NOP- INEA-04.

- The distance between the surface of the water in the pond and the top of the slope crest must not be less than 0.50 metres, even during periods of flooding.
- The internal and external features of the slope should be covered with grasses to prevent erosion.

- If an erosion process occurs, erosion containment measures should be put in place, such as revegetation of the area, construction, etc.

  terraces, laying vegetation cordons, installing ridge channels, laying mulch, minimum cultivation, etc.
- Licensing for aquaculture activities does not authorise the direct or indirect commercialisation of material from pond excavations. In cases where commercialisation is envisaged, the interested party must obtain a document proving ownership of the area from the National Department of Mineral Production (DNPM) or an agreement with the holder of the mining rights to the area, as well as a specific environmental licence for extracting the material from the competent body.
- Physico-chemical analyses of the water collected must be presented, covering at least the parameters pH, dissolved oxygen, ammonia and turbidity, carried out by a professional qualified to carry out these analyses, who presents proof of good standing with the professional council.
- Fish farming enterprises in ponds, lined tanks or raceways must have effluent treatment systems, such as sedimentation basins, mechanical and biological filtration systems and water reuse, or other treatment systems that are equally effective and meet the specifications of CONAMA Resolution No. 357/2005, supplemented and amended by CONAMA Resolution No. 410/2009 and CONAMA Resolution No. 430/2011, and Technical Standard NT-202.R-10, which establishes criteria and standards for the discharge of liquid effluents, a revision approved by CECA Decision No. 1007 of 04 December 1986 and published in the Official Gazette of 12 December 1986.
- The outlet of effluent treatment systems must be protected with a screen or filter to prevent

fish or fry from escaping.

- In the case of sedimentation basins, these should be at least 10 per cent of the total volume of water in the production area, so that the water discharged through the drainage structure remains there long enough for the suspended materials to sediment completely.
- Floating aquatic macrophytes of the genus *Eichornia* or another with a similar ecological function, in the growth phase, and native detritivorous fish that consume organic remains or plankton, should be maintained in sedimentation basins to reduce the amount of organic matter in the water.
- Adult aquatic vegetation that is removed periodically must be properly disposed of in accordance with waste disposal regulations.

— The sedimentation basins must be cleaned every six months and any waste that cannot be reused must be disposed of at suitable sites duly authorised in the licensing process.

— A different basin cleaning frequency may be approved, based on technical justification that proves that the effluent will meet the discharge standards required by current legislation.

— The use of in natura manure to fertilise the water in cultivation ponds will only be approved if measures are taken to control the quality of the cultivation water.

— The nurseries should be emptied and maintained periodically and, if necessary, treatments should be carried out to maintain suitable soil conditions, including the application of correctives and drying by exposure to the sun.

Criteria for environmental licensing of net-pen and cage fish farming according to NOP-INEA-04

— The environmental licensing of fish farming enterprises in artificial environments must comply with the classification and destination of water, as set out in CONAMA Resolution 357/2005, supplemented and amended by CONAMA Resolution 410/2009 and CONAMA Resolution 430/2011.

— Cultivation is not authorised in bodies of water that are proven to have periodic occurrences of harmful algae.

— Cultivation of alien species in net tanks and cages is not authorised, except when the species being reared has already been effectively introduced into the water system used.

— The sum of the cultivation areas may not exceed 1.0 per cent of the total area of fish farming projects in net pens and cages.

— Physico-chemical analyses of the water body must be submitted prior to the installation of the net tanks and cages and at least the parameters turbidity, transparency, pH, dissolved oxygen, ammonia, nitrite and dissolved organic phosphorus must be analysed.

— The number of collection points, as well as their location and collection depth, must be established in such a way as to allow the physical and chemical characterisation of the water body between the production point and a control point.

— The net tanks and cages must be made of resistant materials and receive regular maintenance to prevent them from breaking, and special care must be taken when transporting, repairing, handling and harvesting them to prevent the animals from escaping.

— The installation of net tanks and cages must be observed:

- Positioning perpendicular to the chain;
- A distance of at least 10 metres between the rows of tanks;
- Distance between tanks at least equal to their length;
- A distance of at least 1.5 metres between the bottom of the tank and the bottom of the water body.

— The installation of structures that significantly alter the local hydrodynamics, causing damage such as erosion and siltation, will not be authorised.

— The feed supplied must have the consistency, durability and buoyancy characteristics appropriate to the type of rearing.

— The feed should be administered in a quantity and frequency appropriate to the species being cultivated, so as to avoid excesses in the environment that could affect water quality.

— Water analyses must be presented at the cultivation site, covering at least the parameters turbidity, transparency, pH, dissolved oxygen, ammonia, nitrite and dissolved organic phosphorus, at the following frequencies:

- Developments whose impact is classified as low: every six months.
- Projects whose impact is classified as medium or high: quarterly.

Criteria for environmental licensing of ranching enterprises according to NOP-INEA-04.

- Spinning in ponds must follow the same criteria presented for environmental licensing of fish farming in excavated ponds, lined ponds and raceways.

- The water inlets and outlets of each tank or stall must be independent, preventing water from one passing into another, in order to avoid contamination.

- Rotting food and carcass remains should not be placed in the ranches to attract insects for feeding frogs and imagos.

- If the biological inducer - fly larvae *(Musca domestica)* - *is* used to feed imagos and frogs, the flies, in their various life stages, must be reared in moscaria and places equipped with insect screens and traps that make it impossible for them to escape.

- Ranching enterprises must have effluent treatment systems, such as sedimentation basins, mechanical and biological filtration systems and water reuse, or other treatment systems that are equally effective and meet the specifications of CONAMA Resolution No. 357/2005, supplemented and amended by CONAMA Resolution No. 410/2009 and CONAMA Resolution No. 430/2011, and Technical Standard NT-202.R-10, which establishes criteria and standards for the discharge of liquid effluents, a revision approved by CECA Decision No. 1007 of 04 December 1986 and published in the Official Gazette of 12 December 1986.

- The outlet of the effluent treatment systems should be protected with a screen or filter to prevent tadpoles, imagos and frogs from escaping.

- If the effluent treatment system has a sedimentation basin, it must be at least 20 per cent of the production area, so that the water discharged through the drainage structure remains there long enough for the suspended materials to sediment completely.

- The area of the sedimentation basin can be reduced if the effluent that reaches it first passes through a septic tank to retain solids. These areas must be included in the technical project.

- Floating aquatic macrophytes of the genus Eichornia or another with a similar ecological function should be kept in the sedimentation basins in the growth phase, along with native detritivorous fish that consume organic waste or plankton, in order to reduce the amount of organic matter in the water.

- Adult aquatic vegetation that is removed periodically must be properly disposed of in accordance with waste disposal regulations.

- The sedimentation basins must be cleaned every six months and any waste that cannot be reused disposed of at sites duly authorised in the licensing process.

- A different basin cleaning frequency may be approved, based on technical justification proving that the effluent will meet the discharge standards set out in CONAMA Resolution 357/2005, supplemented and amended by CONAMA Resolution 410/2009 and CONAMA Resolution 430/2011, and Technical Standard NT-202.R-10, which establishes criteria and standards for discharging liquid effluents, a revision approved by CECA Resolution 1007 of 04 December 1986 and published in the Official Gazette of 12 December 1986.

- Injured animals or those showing clinical signs of disease should be immediately separated from healthy animals and kept under observation and treatment prescribed by a veterinarian. In the case of infectious diseases, the cultivation facilities should be isolated and the water disinfected before being released into the environment.

- The facilities must be cleaned and disinfected after each batch is removed, using disinfectant solutions permitted by law or a blowtorch.

11.3. CONCLUSIONS

By following the guidelines set out in this chapter, aquaculture farmers will be duly regularised before the environmental agencies and will be able to carry out their activities in accordance with the relevant legislation.

The benefits of operating a licensed business are not just about complying with the legislation in force: by adopting environmental protection and control measures, the product you obtain is worth more on the market, enabling you to reach selected commercial niches and make your business more profitable.

11.4. BIBLIOGRAPHICAL REFERENCES

STATE ENVIRONMENTAL INSTITUTE (INEA). Operational Standard for Licensing Continental Aquaculture - NOP - INEA - 04 approved by CONEMA Resolution No. 49 of 02/08/2013 and published in the D.O.U. of 14.08.13.

CONAMA - NATIONAL ENVIRONMENTAL COUNCIL. Resolution No. 357, of 17 March 2005.

CONAMA - NATIONAL ENVIRONMENTAL COUNCIL. Resolution No. 430, of May 2011.

FEEMA - STATE ENVIRONMENTAL ENGINEERING FOUNDATION. Technical Standard - Criteria and Standards for the Release of Liquid Effluents - NT-202. R-10. Rio de Janeiro, 1986.

CHAPTER 12

MARKET AND COMMERCIALISATION

Eduardo Shimoda

Rodrigo Martins Pinto

Vitor Gomes Viana da Silva

12.1. INTRODUCTION

Fish meat was used as food by primitive men. From bone and stone fragments and drawings engraved in caves, it can be deduced that primitive men were hunters and fishermen (SANTOS, 1977).

Fish is an excellent source of high-quality, quickly digestible proteins, rich in nutrients such as vitamins A and D, which are not usually found in routinely consumed foods, as well as vitamins B, E and K and phosphorus (LANKE; MAGAGNIN; FURLONG, 2003). Fish meat also contains Omega-3 fatty acids, components that are responsible for proven health benefits, which is why it is advisable to eat at least three fish meals a week. Fish meat is soft and highly digestible and, with the consumption of 150 grams daily, an adult weighing approximately 70kg absorbs the necessary proteins. Although the percentage of protein present in mammalian and poultry meat is similar to that of fish, the latter has protein of greater biological value due to its utilisation (KURIEN, 2005).

Some studies have been carried out to ascertain the profile of consumers of certain types of meat (CANEVER et al., 1997; SILVA; FABRINI FILHO, 1994). Among the objectives for carrying out research analysing the consumers of any product is the adoption of marketing strategies in order to meet the population's needs. According to Engel et al. (2000), consumer behaviour should be studied to understand the activities directly involved in obtaining, consuming and disposing of products, and it is interesting to include decision-making processes before and after the actions.

As for the marketing aspects, the entire production, trade and distribution sector must be concerned with prices and the cost of production, since it is this system that must adapt to the market and not the other way round. Therefore, if price is so important to consumers, practices should be adopted that make it possible to reduce production costs in order to provide a lower-priced and more stable product, given that supply is not regular and prices fluctuate frequently throughout the year. It is therefore crucial to evaluate the entire commercialisation chain, not just the sales points. This will make it possible to develop the activity with the appropriate technique, responsibility and most importantly: sustainability (ARAUJO, 2006).

This chapter presents the perception of the population of Campos dos Goytacazes regarding the market and commercialisation of fish, comparing it to other types of meat.

12.2. METHODOLOGY

The work consisted of a survey, using interviews and questionnaires, of residents of the municipality of Campos dos Goytacazes, from October 2007 to March 2008.

The questions initially focussed on meat consumption habits in general, comparing frequencies of consumption of chicken, fish, beef and pork. Next, the questions focused on fish meat, including, among other things, price assessment, preferences for origin (marine or freshwater). Finally, the interviewees were asked for personal details that would allow them to be characterised according to their level of income, education and age group.

A total of 92 questionnaires were administered, and the sample was stratified and proportional to the population of the municipality of Campos de Goytacazes (IBGE, 2001). The frequencies found were compared using the chi-squared test to check for differences between genders, incomes, education levels and age groups in the various items of the survey.

12.3. RESULTS AND DISCUSSION

12.3.1. Preferred place of purchase

Figure 1 shows the preferred place of purchase for the different types of meat, where beef, poultry and pork are preferably bought in supermarkets. Fish meat is bought more at fairs than at other places.

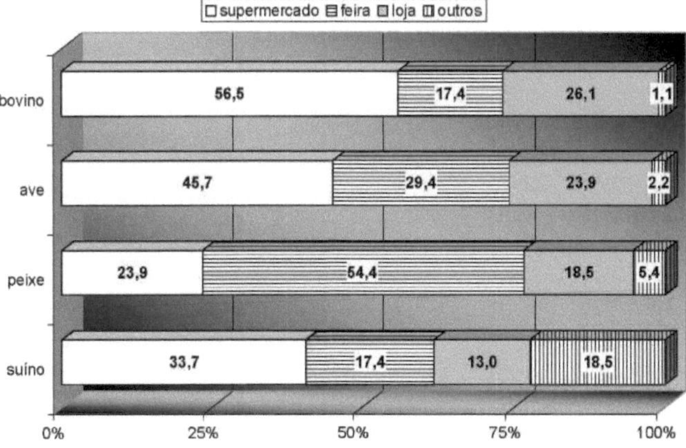

Figure 1- Preference (%) for the place of purchase of the different types of meat.

12.3.2. Consumer evaluation of fish prices

When asked how they rated the price of fish, all the results showed that the majority considered it to be affordable, with an average close to 3.0 (Figure 2), both in the general result and in all the stratifications carried out (gender, income, schooling and age). There was no influence of these classes on price evaluation (P>0.05).

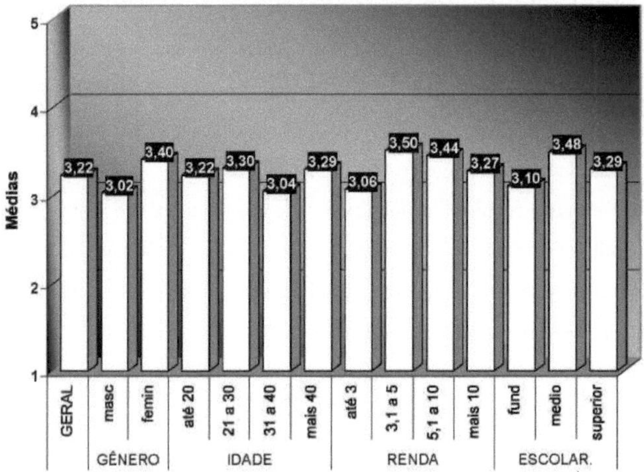

Figure 2- Average score given to the evaluation of the price of fish meat (scale from 1 to 5, where 1 - very cheap to 5 - very expensive).

12.3.3. Acceptable price for fish

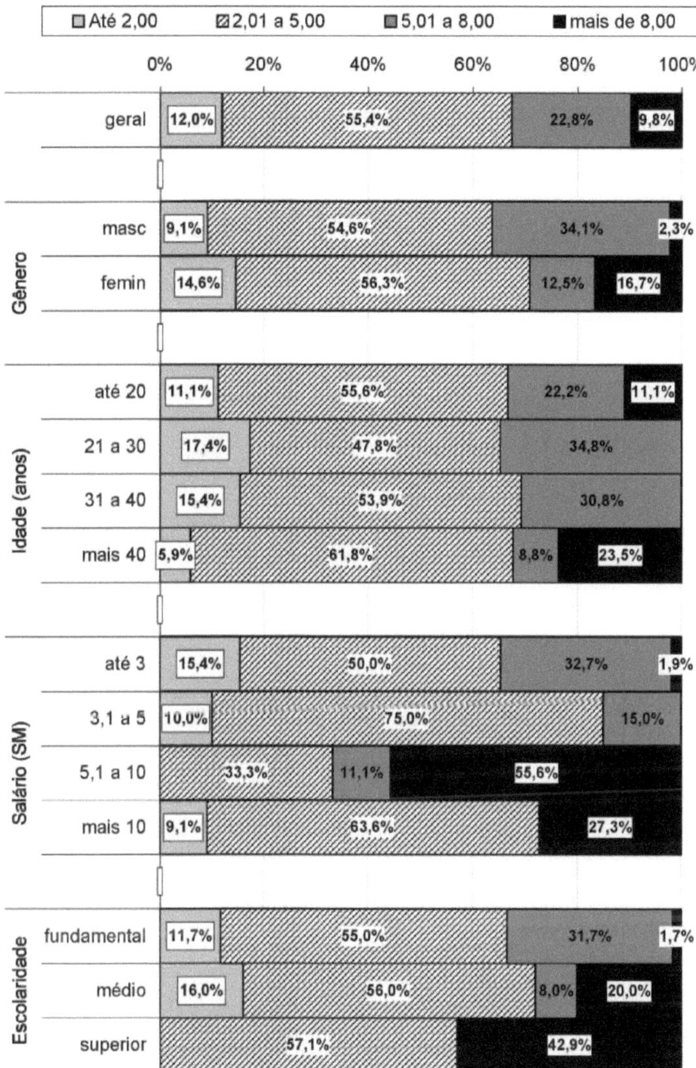

Figure 3 - Price of fish considered acceptable by the interviewees: overall result and according to gender, age (years), income (minimum wages) and schooling.

With regard to gender, there was a significant difference in what people were willing to pay for a kg of fish (Figure 3), with 54.6% of men considering a price between R$2.01 and R$5.00 to be an affordable price
and 56.3 per cent of women shared the same opinion. In the 5.01 to 8.00 price range, there was a greater difference, as 34.1% of men considered this range to be acceptable, while only 12.5% of

women did.

There was a significant difference (P=0.0266) in what was considered acceptable to pay for a kg of fish when comparing different age groups, with interviewees aged over 40 being the stratum in which there was the highest proportion (23.5 per cent) of people willing to pay more than R$8.00. In the other age groups, the percentages who would agree to pay more than this amount ranged from 0 to 11.1 per cent.

As for the income bracket of those interviewed, there was a statistical difference (P<0.0001) as to the price considered acceptable for fish. Of those earning up to 3 minimum wages, only 1.9% considered paying more than 8.00 reais for a kg of fish to be affordable, while 50% considered the price between 2.01 and 5.00 reais to be acceptable. Of those earning between 3.1 and 5 salaries, 75% were more accepting of the price range between R$2.01 and R$5.00. Still in this income bracket, no interviewee accepts buying fish meat when the price is more than R$8.00, while those earning between 5.1 and 10 salaries, 55.6% considered this price to be affordable.

As for the level of education, there was a significant difference (P=0.0013) in the acceptable price for fish meat. The stratum that included people with higher education had the highest percentage (42.9%) of interviewees who agreed to pay more than R$8.00 for a kilogram of fish.

12.2.1. Preference according to the origin of the fish

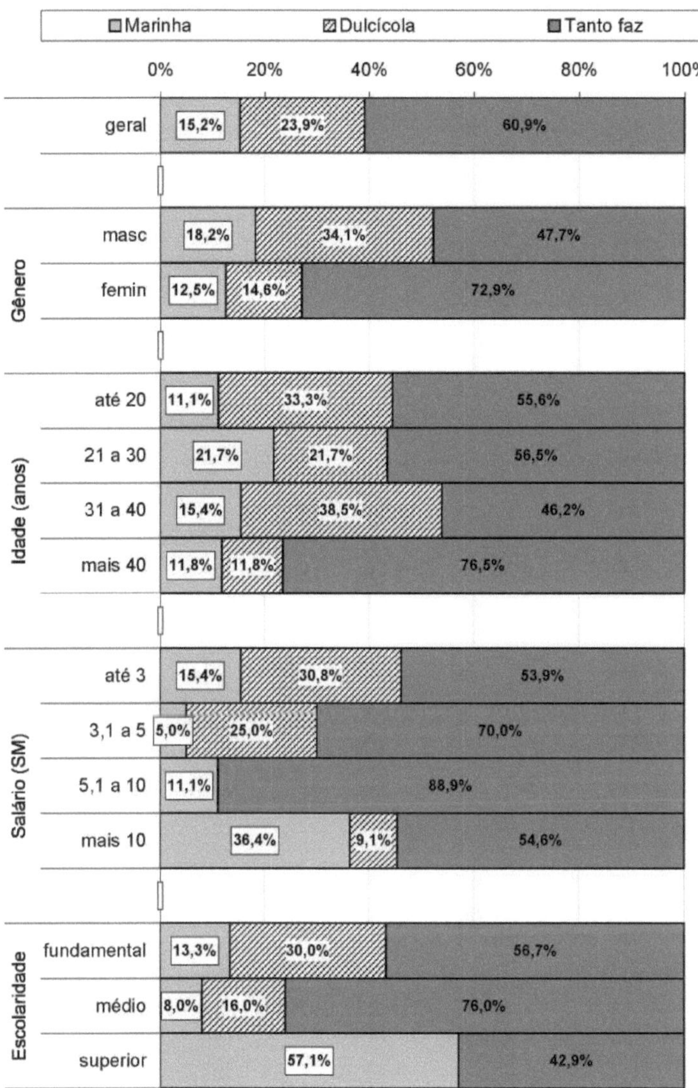

Figura 4-   Interviewees' preference according to the origin of the fish (marine or freshwater), stratified by gender, age (years), income (minimum wages) and schooling.

As for the origin of the fish (Figure 4), 47.7% of the male interviewees had no preference between marine or freshwater fish and 72.9% of the female interviewees shared the same opinion. There was a greater difference when the preference was for freshwater fish, with 34.1% of men and 14.6% of women making this choice.

With regard to age (Figure 4), there was no significant difference (P=0.2248) in preference according to the origin of the fish. In all income brackets, it was found that most people had no preference between sea and freshwater fish, and there was no significant effect of income bracket on preference. On the other hand, schooling had a significant influence (P<0.05) on preference. Interviewees with primary or secondary education showed no preference as to origin, while people with higher education showed a preference for fish from marine waters.When analysing the reason for the interviewees' preference for fish (Figure 5), it was noted that 51.4% of the interviewees like to eat it because of the taste. With regard to marine or freshwater species, it can be seen that in both cases the factor that has the most effect on the choice of source is flavour: 61.5% of those who prefer sea fish cite taste as the reason for their choice, while 45.5% of those who prefer freshwater fish also cite flavour as the reason for their choice.Consumers have shown a remarkable preference for fresh products, frozen portions and industrialised convenience foods, and with the growth of this segment, the food industry has been attracted to invest in the development of semi-prepared products, which reduce the time spent preparing food at home, ready-to-eat products, ready-made or baked, which have shown considerable growth despite the also growing offer of products in fast food chains (LIMA et al. 1995).

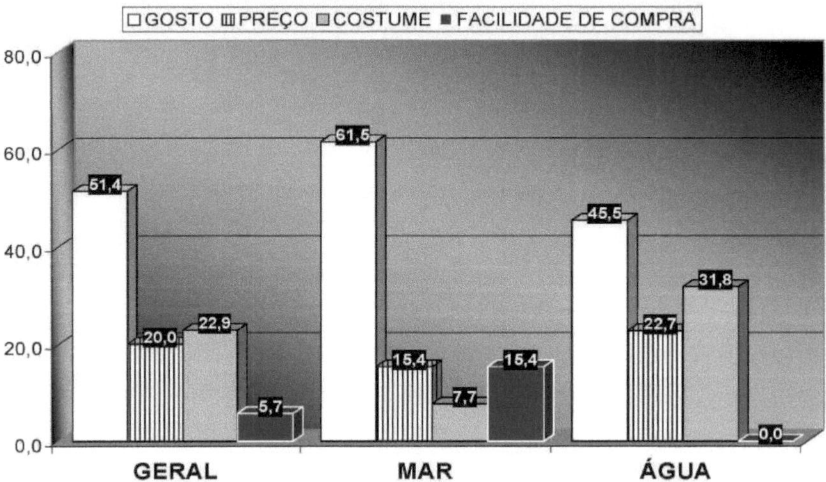

Figure 5 - Criteria adopted by the interviewees for choosing fish of marine or freshwater origin: general results and stratified according to the origin chosen.

12.4. Conclusions

Based on the results of the analyses carried out on the basis of the data collected in the field research in the municipality of Campos dos Goytacazes, it can be concluded that:

1) Unlike other types of meat, which are preferably bought in supermarkets, fish is preferably bought at fairs;

2) The majority of the population of Campos dos Goytacazes considers the price of fish to be reasonable;

3) The best price range for a Kg of fish accepted by the interviewees was between R$2.01 and R$5.00;

4) For most residents of Campos dos Goytacazes, it makes no difference where the fish comes from (fresh or salt water)

REFERENCES

ARAUJO, Marcelo Gomes de. **Socio-economic and technological characterisation of fish farmers in the central region of the state of Tocantins, using multivariate analysis techniques.** 2006. 100 f. Thesis (Doctorate) - Zootechnics Course, Federal University of Lavras, Minas Gerais, 2006.

BARCELLOS, M. D. de; CALLEGARO, C. A. M. **The importance of information as quality indicators: the case of beef purchases in Porto Alegre.** In: XXVII Congresso da Associação Nacional dos Programas de Pós- Graduação em Administração, Anais... 2002, Salvador/BA.

BLEIL, S. I. **The Western dietary pattern: considerations on changing habits in Brazil.** Caderno de Debates UNICAMP, Campinas, v. 6, p. 1-25, 1998.

CANEVER, M. D. et al. **Competitiveness between the Brazilian and Argentinian chicken meat chains.** Brazilian Congress of Rural Economics and Sociology, 35. Natal, 1997. Proceedings... Natal, 1997.

DELGADO, C L ; WADA, N ; ROSENGRANT, M W ; MEIJER, S ; AHMED, M (2003). Outlook for fish to 2020: Meeting global demand. International Food policy Research Institute, WorldFish Centre, Penang, Malaysia: 28p.

ENGEL, J. et al. **Consumer Behaviour.** 8.ed. Rio de Janeiro: LTC, 2000.

GONÇALVES, J. S.; MACHADO, R. S. **Consumo e hierarquia dos relativos de preços de proteína animal no Brasil,** 1997-2006, Informações Econômicas, SP, v.37, n.9, sep. 2207

IBGE. **2000 demographic census:** population and household characteristics. IBGE. 2001.

KURIEN, J. **Responsible fish trade and food security.** FAO Fisheries Technical Paper, No. 456: 102p, 2005.

LANKE, N.G et al. **Determination of vitamin A in fish: adaptation of methodology. Rev. Inst. Adolfo Lutz,** 62(3):151-158, 2003.

LIMA, J. et al. **Poultry farming - Sector report.** 1995.

OLIVEIRA, S. P.; THÉBAUD-MONY, A. **Agro-industrial consumption model: homogenisation or**

diversification of eating habits? Caderno de Debates, Campinas/SP, v. 4, p. 1-13, 1996.

SANTOS, E. **Fishing and fish farming.** Belo Horizonte, Itatiaia Ltda, 1977.

SILVA, L. F.; FABRINI FILHO, L. C. **Poultry complex and questions about eating habits. Caderno de Debate UNICAMP**, Campinas, v. 2, p. 41-61, 1994.

CHAPTER 13

Economic Viability of Tilapia Farming in Ponds

Alcimar das Chagas Ribeiro
Débora da Silva Florenzano
Gabrielle Guimarães Klerch Alves
Poliana Romero Miller
Yasmin Pereira Gonçalves

1. Introduction

According to Vilela (2013), fish farming has become an attractive economic activity, especially for small and medium-sized producers, as it does not require large tracts of land or large investments, and is a promising activity. Fish farming is a practice that contributes to the conservation of certain species, provides quality food for consumers and a considerably high financial return, when appropriate financial planning is applied.

According to Zimmermann and Hasper (2003 apud SILVA, 2007), the most widely cultivated species in Brazil is the Nile tilapia *(Oreochromis niloticus),* as it adapts better to living in ponds and tanks than other types of fish and is more widely accepted on the market due to its desirable characteristics and excellent feed conversion.

All this economic and productive attractiveness has meant that "Brazilian continental **aquaculture** has grown continuously, reaching a production of approximately 394,340 tonnes of fish in 2010, while tilapia farming **reached the level of 155,450.80 tonnes produced"** **(BRASIL, 2012).**

Coracini (2004) points out that São João da Barra is a municipality in which 5% of its population is directly involved in artisanal fishing. However, it can be seen that there are many obstacles, such as pollution in rivers, seas and lakes and overfishing - where a species is taken in large quantities from its natural habitat before it is harvested.

The fact that aquaculture is not even able to reproduce, preventing its natural replenishment and potentially leading to the species' mass extinction - has caused a sharp decline every year and an increase in poverty in these regions. Because of this situation, aquaculture is gaining more and more ground in the region.

In view of the previous experience based on plastic canvas tanks in São João da Barra, some producers have come up with technological innovations, seeking the economic sustainability of commercial fish farming. Among these innovations, the construction of tanks lined with reinforced concrete is worth highlighting, in order to improve the durability of the structure.

This chapter therefore studies the economic viability of tilapia farming using this innovation.

Specifically, this analysis aims to analyse the economic viability of raising tilapia in five concrete tanks for a 180-day cycle, with a maximum productivity per tank of 200 kg per cycle, on a small family property in Balneário de Atafona - São João da Barra - RJ.

2. Material and methods

This section aims to present the methods used to analyse the financial viability of tilapia farming. The production area is organised into five concrete tanks measuring 7 m x 14 m x 1.30 m and with a useful volume of 120 m^3. The species chosen for production was the Nile tilapia, as it adapts to the region's climate and has a moderate rearing cycle compared to other fish species. Its higher productivity and acceptance in the consumer market can also be highlighted.

In the initial production phase, 500 fry were placed in each tank. This number was established in order to favour the natural oxygenation of the fish and not harm their growth. Subsequently, the tilapia were fed - three times a day according to the proportion of the animal's biomass - with feed proportional to their development until they completed a 180-day cycle, when they were already at an average weight of 700g.

The average production achieved was 200kg per tank in each 180-day cycle. Of this total, 60kg is allocated to fillet production. Of the remainder, 60kg is allocated to meatball production and 80kg, which corresponds to the remains of fillet and meatball production, is discarded as it is not suitable for commercialisation.

The study was carried out using exploratory data research, with the help of a literature review and data analysis. The approach used in this study is quantitative, as the conclusions are drawn from the results obtained from the calculations made. The tables used were based on the article by Campos (2007). In order to carry out the economic feasibility analysis, data was collected on the entire process, from initial investment to implementation. In addition, equipment and all the resources needed to build the facilities were considered as investments. The data used was obtained from the producer in 2016 and organised in Excel spreadsheets. The data was divided into the categories of implementation costs and operating costs, which served as a guide for this analysis.

Based on the implementation data, the project's investment value was obtained, i.e. the amount earmarked for the start of the project, which can be considered quite high, but given the useful life of the installation - around 50 years - the investment is justified. The operating costs were used to calculate the project's expenses. The cash flow was then calculated per year based on the initial investment to implement the project and the annual expenditure (outflows) and annual income (inflows).

To calculate the cash flow, the value of income, investment and expenditure over a given period was used. The net flow was considered to be Revenue (Investments + Expenses). The data will be presented in the results section.

The economic viability indicators - Net Present Value (NPV) and Internal Rate of Return (IRR) - were then used to assess the profitability of the project over a 15-year horizon. Net Present Value is understood as follows: firstly, a certain interest rate (also known as the discount rate) is accepted. Therefore, **the net present value** can be defined as the algebraic sum of the cash flow balances discounted at that rate on a given date (WOILER; MATHIAS, 2008). The rates used to calculate the NPV were 10, 12 and 15 per cent. The Internal Rate of Return, considered the discount rate that makes the net present value of the investment zero and indicates whether the project is viable or not.

The higher the IRR, the more guaranteed and viable the investment, or it is enough for the IRR to be higher than the discount (WOILER; MATHIAS, 2008).

3. Results and discussions

The investment implementation data for tilapia farming covered in this study is shown in Table 1. The percentage share of each item is shown in Figure 1.

Table 1 - Investment items and values for tilapia production in 5 concrete tanks, in Real (R$), 2016

Investment	Quantity	Value each (R$)	Total
Equipment			
Building material*	5	7.500	37500
Soil preparation	5	600	3000
Electric water pump	2	400	800
Scales	1	450	450
Trawl net	1	400	600
Freezer	2	1000	2000
Styrofoam tray	1	51,2	51,2
Sealing machine	1	429	429
Installations			
Construction labour	5	6000	30000
Investment cost			**74830**
Building material*= (cement, sand, clay, iron, etc.}			

Source: Own authorship based on the article CAMPOS, 2007

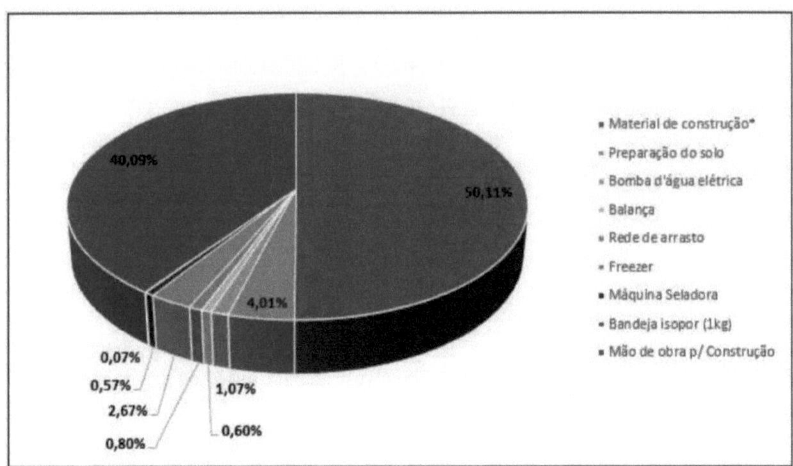

Figure 1 - Share of items in the investment cost of tilapia farming in concrete tanks in the municipality of São João da Barra, RJ. Source: Free elaboration based on data from the investment table.

To calculate costs, labour was taken into account, as well as the inputs needed for tilapia production, the cost of depreciating equipment and facilities, and other project expenses. To do this, the total value per cycle was calculated, followed by the total value per year, which will provide the basis for analysing the cash flow. This resulted in the total operating cost. This is shown in Table 2 below:

Table 2 - Total cost of tilapia production in concrete tanks in the municipality of São João da Barra, RJ

Item	Quantity/cycle	Unit price (R$)	Total value/cycle (R$)	Total value/year (R$)
Labour				
Permanent labour (salary)	1	900,0	900,0	1800,0
Subtotal			900,0	1800,0
Inputs				
Fry (500un.)	5	75,0	375,0	750,0
Feed*	5	500,0	2500,0	5000,0
Styrofoam tray	600	0,128	76,8	153,6
Subtotal			2.951,8	5903,6
Other Expenses				
Energy	6	84,2	505,4	1010,9
Communication	6	50,0	300,0	600,0
Meatball ingredients**	1	30,0	30,0	60,0
Transport	6	50,0	300,0	600,0
Subtotal			1135,4	2270,9
Effective Operating Cost			**2270,9**	**4541,76**
Depreciation per year				
Machinery and equipment (10% p.a.)	1	10479,0	524,0	1048,0
Installations (5 % p.a.)	1	4500,0	112,5	225,0
Subtotal			636,5	1273,0

Total Operating Cost			**5623,7**	**11247,5**

Source: Own authorship based on the article CAMPOS, 2007

According to Table 2, feed ends up becoming a relevant item for the management of the fish production system, representing approximately 51% of the Effective Operating Cost (EOC). Thus, the association has a total operating cost of R$11,247.50 to produce tilapia in 5 concrete tanks with a useful volume of 120 m^3 each, considering an initial stage of 500 fry per tank in the cultivation system.

As this is a family-run production system, the producer eventually spends R$900.00 per cycle on a day labourer, free of social security charges. The other services, such as treating the fish, filleting, making meatballs and so on, are carried out by members of his family. At the end of each cycle, after paying off the investment, the profit is divided equally between each member.

Table 3 shows the cash flow, taking into account the amount invested in equipment and facilities, considering as revenue the value of selling the kilogram (kg) of fillet for R$25.00 and the meatball for R$15.00 per kg, totalling 600 kg of fillet and 600 kg of meatball per year, in an estimated 9 years of production.

Table 3 - Cash flow from tilapia production in 5 concrete tanks, in Real (R$), 2016

Year	INPUT (R$)	OUTPUT (R$)		NET FLOW (R$)
	Revenue (fish sales)	Investment	Operating Expenses	
1	0	74830,2	0,0	74830,2
2	24000		11247,5	-62077,7
3	24000		11247,5	-49325,2
4	24000		11247,5	-365 72,7
5	24000		11247,5	-23820,2
5	24000		11247,5	-11067,7
7	24000		11247,5	1684,3
8	24000		11247,5	14437,3
9	24000		11247,5	27189,8
10	24000		11247,5	39942,3
11	24000		11247,5	52694,8
12	24000		11247,5	65447,3
13	24000		11247,5	78199,8
14	24000		11247,5	90952,3
15	24000		11247,5	103704,8

Source: Own authorship based on the article CAMPOS, 2007

Finally, the feasibility indicators, the internal rate of return (IRR) and the net present value (NPV) were calculated. For the NPV, three different rates were used for the calculation in order to obtain a comparison of the values returned. Table 4 shows the results of these calculations.

Table 4 - Economic viability indicators for tilapia production in 5 concrete tanks in the municipality of São João da Barra, RJ

Economic viability indicator	10%	12%	15%
NPV (R$)	488414,44	409647,17	316634,95

Source: Own authorship based on the article CAMPOS, 2007

It was found that the cost of building materials and labour represents around 90% of the capital invested, making the investment relatively high, given the annual return. However, taking into account that the tanks are made of concrete, a material with a relatively long useful life, the investment becomes proportional (FIGUEIREDO, 2016).

It can be seen that the number of 500 fry placed in the ponds is small compared to those placed in net ponds (SABBAG et al., 2007), with approximately 5 fry being distributed per m^3 of water. This is justified by the fact that oxygenation is natural, and increasing the number of fish per m3 would interfere with their final production. It is clear that the biggest expense in terms of operating costs is the feed, which can be exchanged for a lower value feed without altering the productivity and quality of the product, but this can only be confirmed once a price survey has been carried out.

It is also suggested, as a way of alleviating the problem, that associations or cooperatives be formed, as there is a project in the municipality that supports fishermen (RIBEIRO, 2010), so that they can buy more feed direct from the factory, as a strategic way of joint marketing in the pre-production phase.

The percentage of depreciation of the facilities was considered to be 5% due to the main construction material being concrete, which has a useful life of 50 years according to the ABNT NBR 15575 performance standard.

The economic viability indicators obtained were: IRR= 50.1% and positive NPV varying for each interest rate. The investment will be paid back in the first six years and after that the flow will be positive. One form of improvement that has a direct impact on revenue is better utilisation of the tilapia, as only 30% is used for filleting and 30% for meatballs. There is therefore 40 per cent waste (taking into account the head, skin and tail), which could be repurposed.

According to an article published by the National Society of Agriculture (2015), fish consumption is below that recommended by the World Health Organisation (WHO). An opportune possibility is the incorporation of tilapia into municipal school meals. This is a beneficial alternative both for the producer, who will have a certain destination for his production, and for the municipal schools, which will be promoting the consumption of an important protein in school meals.

4. Conclusions

The economic results were favourable to the activity, and the items that had the biggest impact on the investment were the building materials and labour needed to implement the project, which together totalled 90% of the initial investment. Revenues totalled R$24,000.00/year producing fish and meatballs. The economic indicators - IRR = 50.1%; and positive NPV - show the viability of the investment over a 15-year horizon, as well as presenting the investment factor as the biggest obstacle to a higher return in the short term.

And based on the annual income and expenditure, it can be concluded that this fish farming project is highly profitable for the producer, as he will make more than 50% profit on the income after 6 years (the average time for the producer to pay off the investment). This and the other indices could be improved by making better use of the tilapia and using a lower cost feed, through a strategic form of joint marketing in the pre-production phase.

5. References

ANDRADE, R. L. B.; WAGNER, R. L.; MAHL, I.; MARTINS, R. S. Production costs of tilapia *(Oreochromisniloticus)* on a model farm in the western region of the State of Paraná, Brazil. **Cienc. Rural online,** Santa Maria, v.35, n.1, Jan./Feb. 2005Available in: http://www.scielo.br/scielo.php?script=sci_arttext&pid=S0103-84782005000100032. Accessed May 2017.

BRABO, M. F. 1; PEREIRA, L. F. S.; SANTANA, J. V. M.; CAMPELO, D. A. V.; VERAS, G. C. Current scenario of fish production in the world, in Brazil and in the state of Pará: emphasis on aquaculture. **Acta Fish. Aquat. Res.** 4 (2), 50-58. 2016. Available at: https://seer.ufs.br/index.php/ActaFish/article/download/5457/4607. Accessed in May 2017.

BRAZIL. Ministry of Fisheries and Aquaculture. Fisheries and aquaculture statistical bulletin - 2010. Brasília: MPA, 2012. 129

CAMPOS, C. M.; GANECO, L. N.; CASTELLANI, D.; MARTINS, M. I. E. Avaliação econômica da criação de tilápias em tanque-rede, município de Zacarias, SP.**B. Inst. Pesca**, São Paulo, 33(2), 265 - 271, 2007. Available at: ftp://ftp.sp.gov.br/ftppesca/33_2_265-271.pdf. Accessed in April 2017.

CORACINI, D. A., RIBEIRO, A. C.; ARICA, J.; COSTA, R. L.; ERTHAL, M. Experiência no Cultivo de Tilápias em Tanque revestido com Plástico em Solo Arenoso no Município de São João da Barra - RJ.2004.

SABBAG, O. J.; ROZALES, R. R; TARSITANA, M.A.A; SILVEIRA, A.N. Análise econômica da produção de tilápias (Oreochromisniloticus) em um modelo de propriedade associativista em Ilha

Solteira/SP. **Custos e @gronegócioonline,**São Paulo, v. 3, n. 2, Jul./Dec., 2007. Available at: http://www.custoseagronegocioonline.com.br/numero2v3/analise%20economica.pdf. Accessed April 2017.

SEBRAE (Brazilian Micro and Small Business Support Service. **Raising tilapia in excavated tanks - Natal:** SEBRAE/RN, 2014. 32p. Available at: http://www.bibliotecas.sebrae.com.br/chronus/ARQUIVOS. Accessed in April 2017.

SEBRAE (Brazilian Micro and Small Business Support Service). Opportunities in online retail niches in Brazil: market research series. Available   at: http://www.bibliotecas.sebrae.com.br/chronus/ARQUIVOS_CHRONUS/bds/bds.nsf/1 40bd6c7177d21250a99c4ecd2882ca2/$File/5200.pdf. Accessed on 03/05/2017.

SONODA, D. Y.; SHIROTA, R. Fish consumption in Brazil is below the international average. **Visão agrícola**, n. 11, Jul./Dec., 2012. Available at:http://www.esalq.usp.br/visaoagricola/sites/default/files/va11-mercado-e- consumo01.pdf. Accessed in April 2017.

VILELA, M. C.; ARAÚJO, K.D.; MACHADO, L. S.; MACHADO, M.R.R.Analysis of the economic and financial viability of a fish farming project in excavated tanks. **Custos e @gronegócioonline**, São Pulo, v. 9, n. 3, Jul./Sept., 2013. Available at: http://www.custoseagronegocioonline.com.br/numero3v9/piscicultura.pdf. Accessed May 2017.

WOILER, Samsão; MATHIAS, Washington Franco. **Projects: planning, preparation, analysis.** 2.ed. São Paulo: Atlas, 2008.

Websites consulted:

Fish farming as an alternative for generating work and income. Available at: http://economianortefluminense.blogspot.com.br/2010/07/piscicultura-como- alternativa-de.html

New FAO report shows that fishing and aquaculture production in Brazil is expected to grow by more than 100 per cent by 2025. Available at: http://www.fao.org/brasil/noticias/detail-events/pt/c/423722/.

Economy and Employment. Indicators suggest optimistic outlook for 2017Available at:
http://www.brasil.gov.br/economia-e-emprego/2017/01/indicadores-sugerem-optimistic scenario for 2017.

Brazil will see a 104% increase in fish production thanks to public policies and investment. Available in: http://www4.planalto.gov.br/consea/comunicacao/noticias/2016/brasil-tera-alta-de- 104-fish-production-graces-public-policy-and-investment

Pronaf Mais Alimentos. Credit to increase the production and productivity of family farmers. Banco do Brasil. Available at:http://www.bb.com.br/pbb/pagina- initial/agronegocios/agronegocio---produtos-e-servicos/credito/investir-em-sua- atividade/pronaf-mais-alimentos#/. Accessed in May 2017.

Appliance Consumption Table. Available at: http://www.eflul.com.br/consumidores/tabela-de-consumo. Accessed in May 2017.

Long-Term Interest Rate - TJLP. Receita Federal. Available at: http://idg.receita.fazenda.gov.br/orientacao/tributaria/pagamentos-e- parcelamentos/taxa-de-juros-de-longo-prazo-tjlp. Accessed in May 2017.

Durability of concrete structures is subject to the action of the environment. Available at: https://www.aecweb.com.br/cont/m/rev/durabilidade-de-estruturas-de-

concreto-esta-sujeita-a-acao-do-meio-ambiente_12727_0_0. Accessed in May 2017.

St John's weather   John   da   Barra. Available at:https://pt.climate-data.org/location/33684/ .

Long-Term Interest Rate - TJLP. BNDES (National Bank for Economic and Social Development). Available at: http://www.bndes.gov.br/wps/portal/site/home/financiamento/guia/custos-financeiros/taxa-de-juros-de-longo-prazo-tjlp/.

Fish consumption in Brazil is below that recommended by the World Health Organisation (WHO). National Society of Agriculture. Available at: http://sna.agr.br/consumo-de-pescado- no-brasil-estaabaixo-do-recomendado-pela- oms/.

Study of useful lives for machinery and equipment. Brazilian Institute of Engineering Evaluations and Expertise of São Paulo. Available at:http://www.ibape-sp.org.br/arquivos/ESTUDO_DE_VIDAS_UTEIS.apresentacao.pdf (sealing machine)

CHAPTER 14

PROSPECTS FOR THE SECTOR

Eduardo Shimoda

Rodrigo Martins Pinto

Vitor Gomes Viana da Silva

13.1. INTRODUCTION

There are a number of countries where fish is a source of employment and profit, so that fishing and the fish trade are activities that represent, for many, their main source of income. One example in South America of pioneering fish production and export is Chile, the world's second largest salmon producer, generating around 25,000 direct jobs. World per capita consumption of fish has increased from 11.6kg in 1971 to 15.7kg in 1997, both figures considered low in relation to poultry and beef (DELGADO **et al.**, 2003).

Consumer behaviour and eating habits are changing as a result of the intensification of women's work, the search for healthier food, globalisation, free trade between nations, growing urbanisation and technological innovations in the agricultural sector. If these changes are monitored, it will become clear that consumers are tending to consume healthier foods, including those that are functional and cholesterol-free, as well as foods that are more practical to prepare. According to the National Health Surveillance Agency (ANVISA), functional foods are those that produce metabolic or physiological effects through the action of a nutrient or non-nutrient on growth, development, maintenance and other normal functions of the human body. Omega 3 is an example of such foods, found in large quantities in saltwater and coldwater fish such as tuna, herring, cod, sardines and salmon. It is clear that the trend is to increase the demand for foods such as these fish, which, compared to other meats, are rich in functional foods (BLEIL, 1998; OLIVEIRA; THEBAUD-MONY, 1996; BARCELLOS; CALLEGARO, 2002).

This chapter presents the results of a study aimed at identifying the perception of the population of Campos dos Goytacazes regarding the sector's prospects.

13.2. METHODOLOGY

The work consisted of a survey of residents of the municipality of Campos dos Goytacazes, using interviews and questionnaires from October 2007 to March 2008.

The questions initially focussed on meat consumption habits in general, comparing frequencies of consumption of chicken, fish, beef and pork. Next, the questions focused on fish meat, including, among other things, price assessment, preferences for origin (marine or freshwater), favourite species, as well as fishing habits. Finally, the interviewees were asked for personal details that would allow them to be characterised according to their level of income, education and age group.

A total of 92 questionnaires were administered, and the sample was stratified and proportional to the population of the municipality of Campos de Goytacazes (IBGE, 2001). The frequencies found were compared using the chi-squared test to check for differences between genders, incomes, education levels and age groups in the various items of the survey.

13.3. RESULTS AND DISCUSSION

13.3.1. FISH COMPARED TO OTHER MEATS

It's interesting to note in figure 1 that beef is the most consumed meat, with an average frequency of 3.7 times a week. Poultry is second (2.8 times a week), followed by fish (1.4 times a week). According to Gonçalves and Machado (2007), beef is the most consumed meat in the country, although total consumption has shown a downward trend, falling from 6.3 to 6.0 million tonnes between 1997 and 2005. On the other hand, the same authors describe an increase in consumption of other meats, including fish meat, which rose from 0.9 to 1.1 million tonnes.

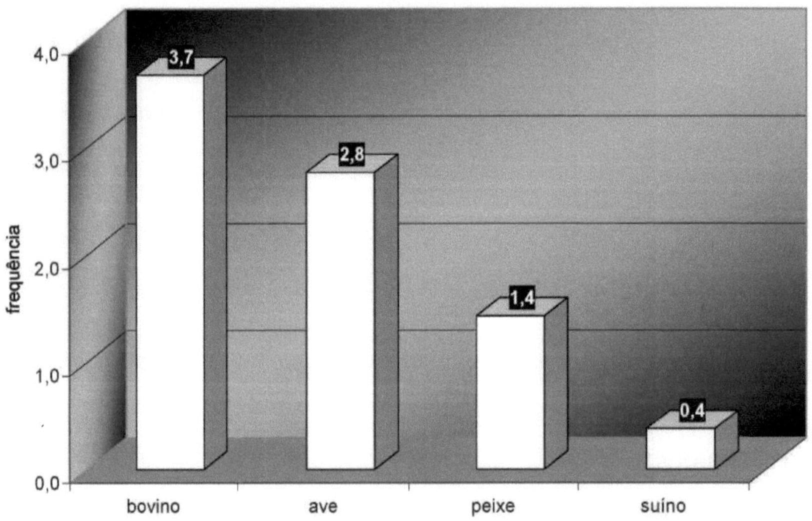

Figure 1- Weekly frequency of consumption by type of meat.

Figure 2 shows how consumers classify each type of meat. The sums are greater than 100 per

106

cent because respondents had the option of answering more than one question. Beef is characterised as healthy and easy to obtain. Poultry and fish, on the other hand, are considered by the majority to be tasty and healthy. Pork is classified as tasty. It should be noted that poultry is considered the cheapest meat, while fish is rated as the tastiest and healthiest of the four meats. This is particularly interesting considering that fish contains high levels of polyunsaturated fatty acids, especially **omega-3s**, which lower blood cholesterol and are recommended for diets where this characteristic is required (KRAUSE; MAHAN, 1991).

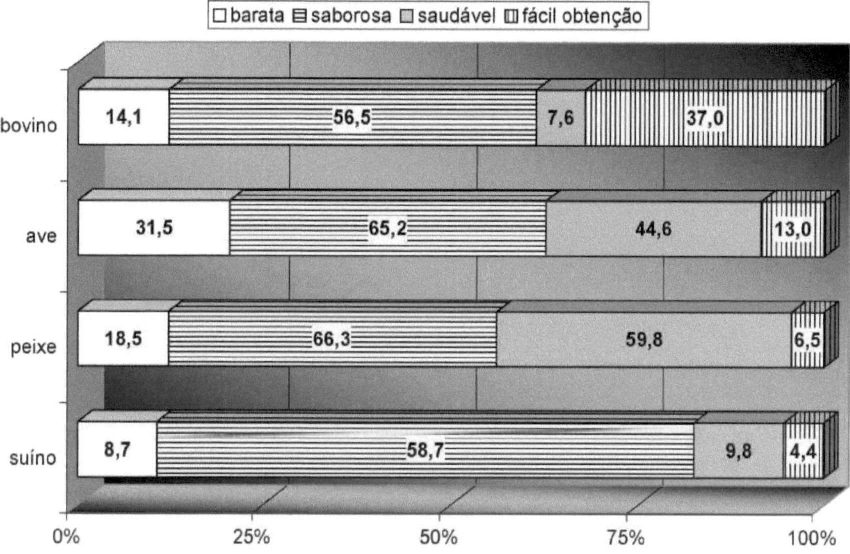

Figure 2- Evaluation of meats in terms of some desirable characteristics.

13.3.2. PEOPLE'S PREFERENCES REGARDING CONSUMPTION AND FISHING

With regard to questions specifically related to fish meat, figure 3 shows that the majority of interviewees prefer to eat fish meat only (52.2%). In addition, 35.9 per cent like to fish and eat it. The proportion of people who said they only like to fish and not eat it was 4.4 per cent.

As for the gender of the interviewees, there was a significant difference (P<0.0001) between the proportions who prefer to only fish, only consume, consume and fish or who don't like to do either of the two. The female interviewees prefer to consume only, while the majority of men like to fish and consume fish.

With regard to age group, there was also a significant difference (P=0.0454) between preferences, with those aged up to 30 predominating and those aged over 30 preferring to eat only fish. It was observed that there was an effect of the income bracket on the interviewees' habits (P=0.0013), and in all brackets there was a higher frequency of those who prefer only to consume fish. The range between 3.1 and 5 minimum wages is noteworthy, with 75 per cent stating that they only prefer to eat fish.

The survey also showed that those interviewed with a primary or secondary education prefer only to consume fish, while those with higher education opted more for the catch and consume alternative, although these differences were not significant (P=0.740).

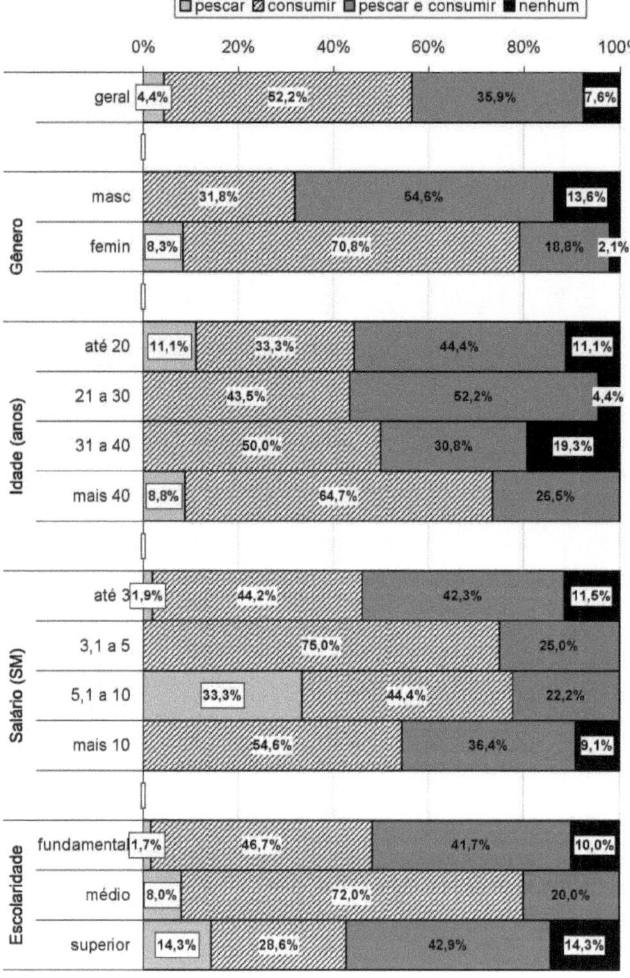

Figura 3- Preference for fishing and consumption, with general results and according to gender (male or female), age group (in years), income (in minimum wages) and schooling.

3.3.4. PEOPLE'S PERCEPTION OF FISH

Respondents were also asked if they thought fish meat was healthy and necessary, and the majority answered in the affirmative, regardless of gender, income, education or age (Figure 4). Although there were no significant differences considering these classifications (all showed P>0.05), the fact that 100 per cent of those earning more than 10 minimum wages or with higher education considered fish meat to be healthy and necessary stands out.

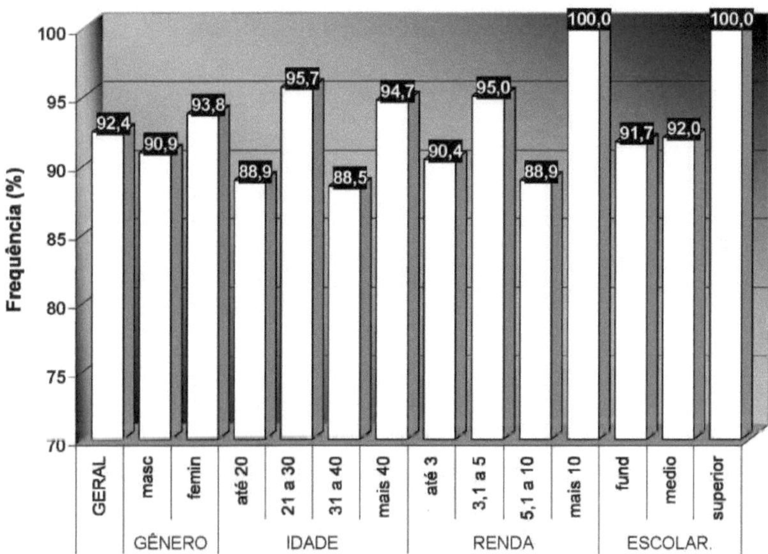

Figura 4-   Proportion of respondents who consider fish meat healthy and necessary.

13.3.5. REQUIREMENTS FOR INCREASING CONSUMPTION

When it was analysed which characteristics of fish meat (price, preparation, ease of finding it) have the greatest effect on a possible increase in consumption, the result was that although 40.8% of those interviewed admitted that they would increase their consumption of fish if it were cheaper, the effect was not significant at the 5% probability level (Figure 5).

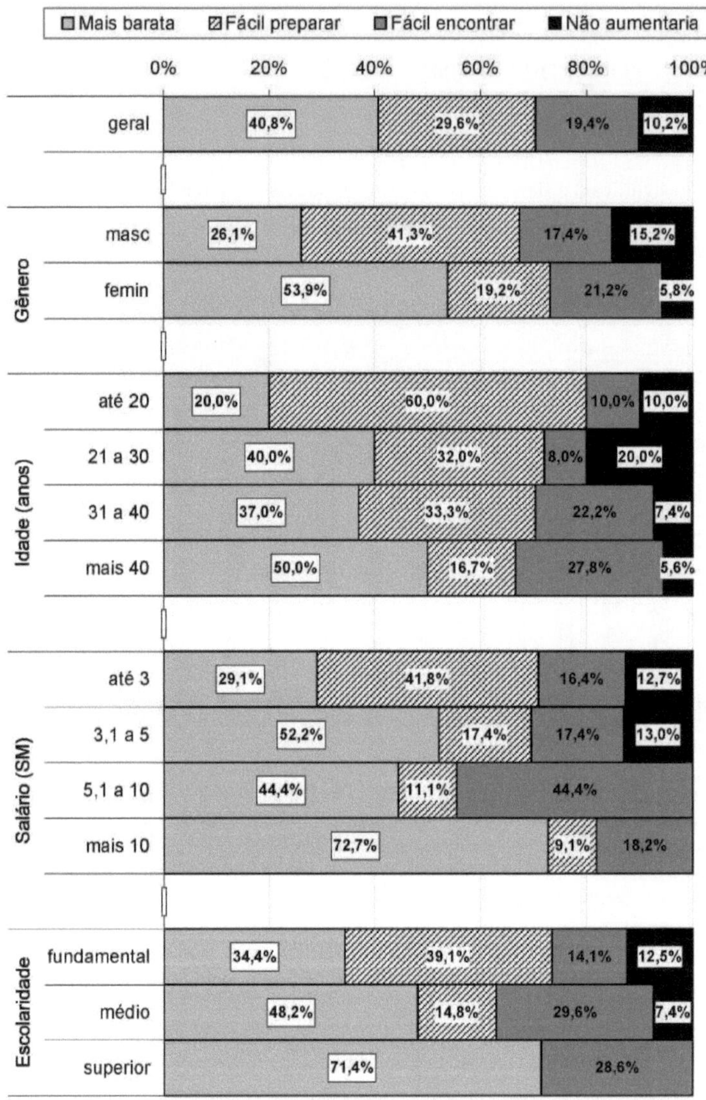

Figura 5- Proportion of respondents who would (or would not) increase their consumption of fish meat by improving the price, ease of preparation or purchase.

With regard to gender (Figure 5), it can be said that there is significance ($P<0.05$) and that most of the men interviewed would increase their consumption if fish were easier to prepare. Women, on the other hand, would increase their consumption of fish mainly if it were cheaper.

When analysing the salary income of those interviewed, significant results were also obtained, bearing in mind that 72.7% of those earning more than 10 minimum wages would increase their consumption if fish were cheaper.

Those earning between 5.1 and 10 minimum wages were divided and said they would increase their consumption both if it were cheaper and easier to prepare. It was also observed that those earning between 3.1 and 5 minimum wages are sensitive to price, as 52.2 per cent of them admitted to increasing consumption if fish became cheaper.

As for age group, there was no significant difference. It was found that of those aged up to 20, 60 per cent would increase consumption if it were easier to prepare, while those aged between 21 and 30 are sensitive to price, seeing it as a condition for increasing consumption. The group aged between 31 and 40, and also those over 40, would increase consumption if it were cheaper, this being the opinion of 37 per cent and 50 per cent of those interviewed, respectively.

There was also no significant influence of level of education (P>0.05). Those with up to primary education said they would increase their consumption both if fish were cheaper and if it were easier to prepare. People with completed secondary education would consume more if fish meat were cheaper. Similarly, 71.4 per cent of those with completed higher education would also increase their consumption if it were cheaper.

13.4 CONCLUSIONS

Based on the results of the analyses carried out on the basis of the data collected in the field research in the municipality of Campos dos Goytacazes, it can be concluded that:

1) Fish meat is only the 3ª most frequently consumed, although it has the potential to grow as it is considered tasty and healthy.

2) More than 90 per cent of those interviewed consider fish to be a healthy and necessary meat to consume;

3) Women prefer only to eat fish, while men like both fishing and eating fish. The older the age group, the greater the preference only for eating fish, with no preference for fishing;

4) Consumption of fish meat would increase for women if it were less expensive, while it would increase for men if it were easier to prepare.

REFERENCES

ARAUJO, Marcelo Gomes de. **Socio-economic and technological characterisation of fish farmers in the central region of the state of Tocantins, using multivariate analysis techniques.** 2006. 100 f. Thesis (Doctorate) - Zootechnics Course, Federal University of Lavras, Minas Gerais, 2006.

BARCELLOS, M. D. de; CALLEGARO, C. A. M. **The importance of information as quality indicators: the case of beef purchases in Porto Alegre.** In: XXVII Congresso da Associação Nacional dos Programas de Pós- Graduação em Administração, Anais... 2002, Salvador/BA.

BLEIL, S. I. **The Western dietary pattern: considerations on changing habits in Brazil.** Caderno de

Debates UNICAMP, Campinas, v. 6, p. 1-25, 1998.

CANEVER, M. D. et al. **Competitiveness between the Brazilian and Argentinian chicken meat chains.** Brazilian Congress of Rural Economics and Sociology, 35. Natal, 1997. Proceedings... Natal, 1997.

DELGADO, C L ; WADA, N ; ROSENGRANT, M W ; MEIJER, S ; AHMED, M (2003). Outlook for fish to 2020: Meeting global demand. International Food policy Research Institute, WorldFish Centre, Penang, Malaysia: 28p.

ENGEL, J. et al. **Consumer Behaviour.** 8.ed. Rio de Janeiro: LTC, 2000.

GONÇALVES, J. S.; MACHADO, R. S. **Consumo e hierarquia dos relativos de preços de proteína animal no Brasil,** 1997-2006, Informações Econômicas, SP, v.37, n.9, sep. 2207

IBGE. **2000 demographic census:** characteristics of the population and households. IBGE. 2001.

KURIEN, J. **Responsible fish trade and food security.** FAO Fisheries Technical Paper, No. 456: 102p, 2005.

LANKE, N.G et al. **Determination of vitamin A in fish: adaptation of methodology. Rev. Inst. Adolfo Lutz,** 62(3):151-158, 2003.

LIMA, J. et al. **Poultry farming - Sector report.** 1995.

OLIVEIRA, S. P.; THÉBAUD-MONY, A. **Agro-industrial consumption model: homogenisation or diversification of eating habits?** Caderno de Debates, Campinas/SP, v. 4, p. 1-13, 1996.

SANTOS, E. **Fishing and fish farming.** Belo Horizonte, Itatiaia Ltda, 1977.

SILVA, L. F.; FABRINI FILHO, L. C. **The poultry complex and questions about eating habits. habits. Caderno de Debate UNICAMP,** Campinas, v. 2, p. 41-61, 1994.

Printed by Books on Demand GmbH, Norderstedt / Germany